A Buddhist Response to the Climate Emergency

Publisher's Acknowledgment

The publisher gratefully acknowledges the generous help of the Hershey Family Foundation in sponsoring the publication of this book.

A Buddhist Response to

THE CLIMATE EMERGENCY

Edited by John Stanley, Ph.D., David R. Loy, Ph.D., and Gyurme Dorje, Ph.D.

Wisdom Publications • Boston

Contributors:

Dalai Lama XIV
Gyalwang Karmapa XVII
Sakya Trizin Rinpoche
Dudjom Rinpoche
Chatral Rinpoche
Thrangu Rinpoche
Dzongsar Khyentse Rinpoche
Ato Rinpoche
Ringu Tulku Rinpoche
Chokyi Nyima Rinpoche
Tsoknyi Rinpoche
Dzigar Kongtrul Rinpoche
Bhikkhu Bodhi
Robert Aitken
Joanna Macy
Joseph Goldstein
Taigen Dan Leighton
Susan Murphy
Matthieu Ricard
Hozan Alan Senauke
Lin Jensen
Thich Nhat Hanh

Information, Quarterly Review, and Blog:

www.ecobuddhism.org

Khor Yug (Environment)
Dedicated calligraphy by Thrangu Rinpoche

Wisdom Publications
199 Elm Street
Somerville MA 02144 USA
www.wisdompubs.org

Library of Congress Cataloging-in-Publication Data

A Buddhist response to the climate emergency / edited by John Stanley, David R. Loy, and Gyurme Dorje.
p. cm.
Includes bibliographical references.
ISBN 0-86171-605-1 (pbk. : alk. paper)
1. Global environmental change—Moral and ethical aspects. 2. Buddhism—Doctrines. I. Stanley, John, 1950– II. Loy, David, 1947– III. Gyurme Dorje.
BQ4570.E23B85 2009
294.3'377—dc22
2009008866

13 12 11 10 09
5 4 3 2 1

Cover design by Gopa&Ted2. Interior design by TLLC. Set in Bembo 12/16.

Wisdom Publications' books are printed on acid-free paper and meet the guidelines for permanence and durability of the Production Guidelines for Book Longevity of the Council on Library Resources.

Printed in the United States of America.

This book was produced with environmental mindfulness. We have elected to print this title on 30% PCW recycled paper. As a result, we have saved the following resources: 24 trees, 17 million BTUs of energy, 2,100 lbs. of greenhouse gases, 8,715 gallons of water, and 1,119 lbs. of solid waste. For more information, please visit our website, www.wisdompubs.org. This paper is also FSC certified. For more information, please visit www.fscus.org.

Table of Contents

Acknowledgments

This book presents the efforts and aspirations of eminent Buddhist teachers, and we are tremendously grateful to all of them. Its story began in late 2006, when Khenchen Thrangu Rinpoche composed a special aspirational prayer, and encouraged us to prepare and publish a pan-Buddhist response to the threat of global warming, together with an accurate account of the relevant science and the solutions. Shortly afterward, we discussed the subject with Kyabje Dudjom Rinpoche in Bodh Gaya. He composed an aspirational prayer, gave key guidance, and suggested the construction of a website as a first step.

In the course of 2007, Ringu Tulku Rinpoche generously provided interviews and he has continued to offer sterling support throughout the project. Interviews and encouragement were provided by Chokyi Nyima Rinpoche, Tsikey Chokling Rinpoche, and Tsoknyi Rinpoche. Through the assistance of Jigme Tenzing Lama, Dzongsar Khyentse Rinpoche composed an aspirational prayer to protect the global environment. Through the good offices of Kunzang Lama of Darjeeling, a prayer was received from the revered Kyabje Chatral Rinpoche. Rinchen Ato arranged a key interview in Cambridge with her father, Ato Rinpoche. Dzigar Kongtrul Rinpoche discussed the subject with us in detail during his visit to Ireland. A profound aspirational prayer was received from the forty-first head of the Sakya tradition, Kyabje

Sakya Trizin. This and the other *monlams* were expertly translated by our co-editor, Gyurme Dorje.

On the Tibetan New Year, 2008, thanks to the essential help of Alexander Venegas, we launched a website: Ecological Buddhism, at www.ecobuddhism.org. It continues to attract significant attention from the international Buddhist community. Buddhist philosopher David Loy then joined us as a co-editor, opening up the project to the essential participation of Zen and other lineages. Important and moving contributions were received from senior American Zen master Robert Aitken Roshi, and from Joanna Macy, a respected leader in the Buddhist movement for peace, justice, and deep ecology. Eddie Stack capably assisted us to produce an intermediate, digitally published version of the book.

In November 2008, we met with the Dalai Lama in Dharamsala to brief him on the climate emergency, especially the rapid degradation of the great glaciers of the Tibetan plateau, upon which depend the major rivers and water supply of Asia. Through the assistance of Ringu Tulku, we were also able to discuss these issues with Gyalwang Karmapa. Meeting these two great spiritual leaders represented an auspicious turning point, at which this project entered upon a more universal stage. It is therefore most appropriate, as well as a great honor, that the Dalai Lama contributes the opening chapter of this book, while Gyalwang Karmapa's essay opens the section entitled "Asian Buddhist Perspectives." As it happens, this section also re-unites four great reincarnate lamas and lineage heads who, between 1959 and 1981, led Tibetan Buddhism into the world beyond the "Land of Snows"—the Dalai Lama, Gyalwang Karmapa, Sakya Trizin, and Dudjom Rinpoche.

We are delighted that other accomplished Western Buddhist teachers then offered their distinctive contributions to the book. In our lifetime, transmission, dedication, scholarship, and practice have made Buddhism a truly global religion. This is fully evident in the originality and authenticity of essays we received from Bhikkhu Bodhi and Joseph Goldstein (representing the Theravada and Insight Meditation traditions), Taigen Dan Leighton, Susan Murphy, Alan Senauke, and Lin Jensen (representing the Zen tradition), and Matthieu Ricard (a Western teacher of Tibetan Buddhism).

It is wonderful that we are able to end this work with a characteristically eloquent essay by Thich Nhat Hanh. The book thereby begins and concludes with the views of two Buddhist leaders who have brought the universal essence of the Buddha's teachings right into the heart of the contemporary world.

We are grateful to Josh Bartok, Laura Cunningham, and Tony Lulek of Wisdom Publications who, understanding the spiritual and practical importance of a Buddhist response to the climate crisis, contributed much personal effort to this book.

The book is respectfully dedicated:

To the human generations to come,
To our companions, the kingdom of the animals of
land, sea, and air,
The great plant kingdom that sustains us all,
The single-celled ones and fungi that recycle and
transform,
All of you who have no voice,
With whom we are inextricably linked
In this net of exquisite energy, this living world.

John and Diane Stanley
Galway, Ireland

Introduction

David R. Loy and John Stanley

The Buddhadharma and the Planetary Crisis

If we continue abusing the Earth this way, there is no doubt that our civilization will be destroyed. This turnaround takes enlightenment, awakening. The Buddha attained individual awakening. Now we need a collective enlightenment to stop this course of destruction. Civilization is going to end if we continue to drown in the competition for power, fame, sex, and profit.

—*Thich Nhat Hanh*[1]

Today we live in a time of great crisis, confronted by the gravest challenge that humanity has ever faced: the ecological consequences of our own collective karma. Scientists have established, beyond any reasonable doubt, that human activity is triggering environmental breakdown on a planetary scale. Global warming, in particular, is happening very much faster than previously predicted.

An increasing number of senior scientists concur with Thich Nhat Hanh that the survival of human civilization, perhaps even of the human species, is now at stake. We have reached a critical juncture in our biological and social evolution. What role might Buddhism play in our response to this predicament? Can the Buddhist traditions help us meet this challenge successfully? These urgent questions can no longer be evaded.

> Our physical environment is changing at a rate that is faster than at any time in the past hundreds of millions of years, except for those rare cataclysmic events that have killed off most life on Earth.
>
> —*Ken Caldeira, Stanford University*[2]

We do not like to think about this ecological crisis, any more than we like to think about our personal mortality. Both individually and collectively, humanity's main reaction has been denial. We repress what we know to be happening—but repression carries a high price. Haunted by vague dread, we become ever more obsessed with competition for profit, power, fame, and sex. Many psychologists believe that people in advanced industrial societies are psychically numbed as a result of being cut off from nature and are unable to feel the beauty of the world—or respond to its distress. The pervasive influence of advertising promises to fill this void, and we spend our time pursuing commodified substitutes that never satisfy. But you can never get enough of what you don't really want.

Escaping this attention trap requires conscious choices based on greater awareness of our true situation. As eco-philosopher and Buddhist scholar Joanna Macy says, denial of what is happening is itself the greatest danger we face. Unfortunately, our collective tendency to denial has been strongly reinforced and manipulated by economic and political forces, whose well-financed advertising and public relations campaigns have succeeded in muddling the issue of climate change.

In June 2008, James Hansen, director of NASA's Goddard Institute for Space Studies and one of the world's most respected climatologists, called for the chief executives of large fossil fuel companies to be put on trial for crimes against humanity and nature. Twenty years ago Hansen's groundbreaking speech to the U.S. Congress warned about the grave dangers of global warming due to human use of carbon fuels. Since then the global climate crisis has become much worse: carbon gas emissions have been increasing radically. If present trends continue, carbon dioxide levels will double by mid-century. Radical steps need to be taken immediately if runaway warming is to be avoided. Not only must we reduce our greenhouse emissions, we need to find ways to remove a lot of carbon dioxide already in the atmosphere.

A dramatic example of our predicament is provided at the poles, where global warming is occurring most rapidly. For far longer than our species has lived on the Earth, the Arctic Ocean has been covered by an area of ice as large as Australia. Now, due to rising air and ocean temperatures, it has been melting rapidly. In 2007, the Intergovernmental Panel on Climate Change (IPCC) forecast that the Arctic might be free of summer sea ice by 2100. It is now apparent that it will disappear within five years. Without the albedo effect of that white ice reflecting the sun's rays, the Arctic ocean will absorb even more solar radiation. This will accelerate the warming of Greenland, whose massive glacial ice sheet alone will, if it melts fully, raise sea levels worldwide by 7 meters.

> The Arctic is often cited as the canary in the coal mine for climate warming... and now the canary has died.
>
> —*Jay Zwally, NASA glaciologist*[2]

Another area critically endangered by global warming is the Tibetan plateau. The mountain ranges that ring it are the source for rivers that supply water to almost half the world's population: the Ganges, Indus, Brahmaputra, Salween, Mekong, Yangtze, and Yellow Rivers, among others. Mountain glaciers maintain those river systems by accumulating ice in the winter and melting slowly in the summer. Himalayan glaciers are receding faster than in any other part of the world, and may well disappear by 2035 or sooner, if the Earth keeps warming at the present rate.

> Now the adverse effects on forests through over-population and the development of various chemical elements in the atmosphere have led to irregular rainfall and global warming. This global warming has brought changes in climate, including making perennial snow mountains melt, thereby adversely affecting not only human beings but also other living species. Older people say that these mountains were covered with thick snow when they were young and that the snows are getting sparser, which may be an indication of the end of the world. The harmful effect on the atmosphere brought about by emissions in industrialized countries is a very dangerous sign.
>
> —*Dalai Lama XIV*[3]

Extreme weather events have quadrupled in frequency since the 1950s. The planetary hydrological cycle has been destabilized, producing bizarre flooding in some places and expanding desertification elsewhere. For the last thirty years, however, our corporations, politicians, and the media they largely control—an "unholy trinity"—have actively resisted the mounting scientific data on the causes and consequences of global warming. What has been the corporate and governmental response to the sudden disappearance of Arctic ice? Oil companies are excited about the prospect of accessing new oilfields. Nation-states are jockeying to claim possession of new territories whose fossil fuel and mineral resources will soon be available for exploitation. This behavior reveals the gap between the economic and political systems we have and the ones we need. Ecologically, such reactions are no less crazy than the alcoholic who thinks that the solution to his hangover is another stiff drink in the morning.

Global warming is one of a number of ecological crises, yet it plays a major role in most of the others—for example, in the disappearance of many of the plant and animal species that share this Earth with us. Almost all scientists are agreed that the Earth is now experiencing a massive species extinction event, the sixth to have occurred in its geological history. Edward O. Wilson of Harvard University, one of the world's most respected biologists, is among those who predict that business-as-usual will drive up to one half of all species to extinction within this century. For a Buddhist, this may be the most sobering statistic of all. What does it mean for a bodhisattva, who vows to save all sentient beings, when most of them are being driven to extinction by our economic and technological activity?

The effectiveness of corporate misinformation about global warming suggests that the critical element of our predicament is lack of awareness—which brings us back to Buddhism. The Buddhist path is about awakening from our delusions. Today we need a collective awakening from collective delusions—including those skillfully manipulated by fossil fuel corporations at a cost of hundreds of millions of dollars a year. The mainstream media—our collective nervous system, so to speak—are corporations whose primary concern is advertising revenue, rather than elucidating for us what is happening to the Earth. We cannot simply rely upon current economic and political systems to solve the problem, because to a large extent they themselves are the problem.

This implies an increasingly urgent need for Buddhists to reflect upon our ecological predicament and bring to bear the resources of our great traditions. The environmental crisis is also a crisis for Buddhism, because Buddhism is the religion most directly concerned with the relationship between the alleviation of delusion and the alleviation of suffering—the *dukkha* (suffering) of all living beings, not just humans. This means that Buddhism has something distinctive to contribute at this crucial time when humanity needs to marshal the best of what it has learned over the course of its history.

The kind of consumer society we take for granted today is so toxic to the environment that continuing business-as-usual is a grave threat to our survival. To address our obsession with consumerism we need different perspectives that open up other possibilities. New technologies alone cannot save us without a new worldview, one that replaces our present emphasis on never-ending economic and technological growth with a focus on healing the relationship between our species and the Earth.

What, more precisely, does Buddhism have to contribute to this urgent conversation? Its traditional teachings offer no easy solution to our environmental crisis, but their familiar critique of greed, ill will, and the delusion of a separate self—the three poisons, which today function institutionally as well as personally—point us in the right direction. Moreover, Buddhism's emphasis on impermanence, interdependence, and non-self implies an insightful diagnosis of the roots of our quandary. To a large extent, our ecological situation today is a greater and more fateful version of the perennial human predicament. Collectively as well as individually, we suffer from a sense of self that feels disconnected from other people, and from the Earth itself.

In contemporary terms, the personal sense of self is a psychological and social construct, without any self-existence or reality of its own. The basic problem with this self is its delusive sense of duality. The construction of a separate self inside alienates me from a supposedly external world outside that is different from me. What is special about the Buddhist perspective is its emphasis on the *dukkha* built into this situation. This feeling of separation is uncomfortable because a delusive, insubstantial self is inherently insecure. In response, we become obsessed with things that (we hope) will give us control over our situation, especially the competition for power, profit, sex, and fame. Ironically, these preoccupations usually reinforce our problematic sense of separation.

The Buddhist solution is not to get rid of the self, which cannot be done since there is no inherently existing self. As Thich Nhat Hanh puts it,

> We are here to awaken from the illusion of our separateness.

When I realize that "I" am what the whole world is doing, right here and now, then taking care of "others" becomes as natural and spontaneous as taking care of my own leg. This is the vital link between wisdom and compassion. My own well-being cannot be distinguished from the well-being of others.

Doesn't this account of our individual predicament correspond precisely to our ecological predicament today? The larger duality is between humanity and the rest of the biosphere, between our collective sense-of-self inside and the natural world believed to be outside. Human civilization is a collective construction which induces a collective sense of separation from the natural world—a sense of alienation that causes *dukkha*. The parallel continues: our response to that alienation has been a collective obsession with securing or grounding ourselves technologically and economically. Ironically (again), no matter how much we consume or dominate nature, it can never be enough. The basic problem is not insufficient wealth or power, but the alienation we feel from the Earth. We cannot "return to nature" because we have never actually been able to leave it. We need to wake up and realize that the Earth is our mother as well as our home—and that in this case the umbilical cord binding us to her can never be severed. If the Earth becomes sick, we become sick. If the Earth dies, we die.

Such a realization implies much greater sensitivity to what is happening to the biosphere, and an acknowledgment of the limitations of human knowledge. We cannot "control" a world that is far more complex than our abilities to understand it. Robert Jensen has called for "the intellectual humility we will need if we are to survive the often toxic effects of our own cleverness."[4]

Our present economic and technological relationships with the rest of the biosphere are unsustainable. We must be radical to be conservative—to conserve what we are currently destroying through exploitation. To survive and thrive through the rough transitions ahead, our lifestyles and expectations must be downsized. This involves new habits as well as new values. Here the traditional Buddhist emphasis on nonattachment and simplicity becomes very important in helping us rediscover and revalue the virtue of personal sacrifice.

> We will need to recover a deep sense of community that has disappeared from many of our lives. This means abandoning a sense of ourselves as consumption machines, which the contemporary culture promotes, and deepening our notions of what it means to be humans in search of meaning.
>
> —*Robert Jensen*[4]

To recognize the seriousness of our situation is to live with a profound sense of grief for what we have collectively done and continue to do. We must not close our eyes to a real possibility raised by recent scientific studies: the extinction of our own species. We are challenged by a new type of *dukkha* that previous generations of Buddhists never faced. Acknowledging this *dukkha* helps us to let go of the delusive competition that distracts us, to focus instead on the crucial work that needs to be done. This grief does not negate the joy of life, which becomes even more precious in light of our heightened awareness of its impermanence. Devotion to doing this great work together can also become a source of great joy. We need new kinds of bodhisattvas, who vow to save not only individual beings but also all the suffering species of a threatened biosphere.

We are challenged now as Buddhists to work together and learn from each other, in order to respond appropriately. By clarifying the essential Dharma of the Buddha, inherent in its diverse cultural forms, we can strengthen its core message for this pivotal time and global society. Although all religious institutions tend to be conservative, Buddhist emphasis on impermanence and insubstantiality implies an openness and receptivity to new possibilities that we certainly need now.

Most people still get their worldview from their religion, and this implies a special responsibility for religions today. If religious worldviews need to be updated in response to the climate emergency, different religions need to do a better job talking to each other and learning from each other. But how can they do that unless different groups within each religion communicate better? What an inspiring example Buddhism could provide, if the various Buddhist traditions were able to work out a joint response to this climate emergency. Given the failure of our economic and political systems, this is an opportunity for religions to rise to the challenge in a way that no other human institutions seem able to do. In this time of great need, the Earth calls out to us. If Buddhism does not help us hear its cries, or cannot help us respond to them, then perhaps Buddhism is not the religion that the world needs today.

> Ah, World! It's in your lap we do our lives and deaths
> It's on you we play out our pleasures and pains.
> You are such a very old home of ours;
> We treasure and hold you dear forever.
> We wish to transform you into the pure realm of our dreams,
> Into an unprejudiced land where all creatures are equal.
> We wish to transform you into a loving, warm, gentle goddess.
> We wish so very firmly to embrace you.
> To that end, be the ground which sustains us all.
> Do not show us the storms of your nature's dark side,
> And we, too, will transform you, all your corners,
> Into fertile fields of peace and happiness.

May the harvest of joyfulness and freedom's million
sweet scents
Fulfill our limitless, infinite wishes, so we pray.

—*Gyalwang Karmapa XVII*[5]

Overview of the Book

The book begins and concludes with contributions (Parts I and VI) from the two most influential Buddhist teachers of our times: the fourteenth Dalai Lama, Tenzin Gyatso, and the Vietnamese master Thich Nhat Hanh. As their essays reveal, the climate emergency has become a paramount concern for both of them.

Part II provides a summary of the most recent scientific findings on the climate crisis, as well as related developments across the spectrum of environment and energy. The information is presented in a broad historical-evolutionary context, which incorporates a Buddhist perspective on how our present situation developed.

The following two sections form the heart of this collection. They offer a variety of Buddhist perspectives on the climate and sustainability crisis, by many well-known Asian (Part III) and Western (Part IV) Buddhist teachers. The first section opens with an essay by Gyalwang Karmapa and includes several aspirational prayers (monlam) by eminent Tibetan masters. Since this type of Buddhist meditative prayer may be less familiar outside Asia, Gyurme Dorje offers an explanatory preface to the section (footnotes to this section, and the translations, precede the general references on p. 279). A functional division "Asian Buddhist Perspectives" and "Western Buddhist Perspectives" was made to structure the book. The intention here is not to create an artificial division in the one world of Buddhism, but to acknowledge how Buddhism has been transmitted in our time, and also to show how the new, global, Buddhist world is coming together over this crucial issue.

Steadily increasing awareness of the global ecological crisis and its implications means that these two parts could have been expanded indefinitely, with contributions from many other Buddhist figures. While we regret the absence of other Asian traditions in Part III, considerations of time and space placed limits on what could be included. We believe that it is important to publish this volume in 2009, the year when the United Nations Climate Conference in Copenhagen will decide a successor to the Kyoto treaty.

Part V reviews some major, collective responses we urgently need to implement if we are to manage and reverse the climate emergency. These solutions have intellectual, psychological, and social as well as technological dimensions. The emphasis is on scientific validity, proven efficacy, the absence of side-effects, and consistency with Buddhist values. In other words, here are solutions that work, that we can begin to actualize now. The section is not meant to be exhaustive, but to outline the direction of a rapid positive transformation of global society. We must all inform ourselves and play our part to assure a safe-climate future, for human civilization and for all the other beings who share this precious world with us.

JUST AS A MOTHER, even at the risk of her own life, protects her child, her only child, in the same way one should cultivate a measureless heart of love toward all beings. One should cultivate a measureless heart of love toward the whole world—above, below, and across—without constriction, enmity, or rivalry. Whether standing, walking, sitting, or lying down, as long as one is awake, one should maintain this mindfulness: they call this "divine living" in this world.

—*Suttanipata*

Part I
Mind, Heart, and Nature
The Fourteenth Dalai Lama

Dalai Lama XIV, Tenzin Gyatso (b. 1935), is the revered spiritual leader of the Tibetan people and former temporal ruler of Tibet. Born in Takster, in Amdo (northeastern Tibet), into a peasant family, he was recognized as the fourteenth Dalai Lama, the embodiment of compassion, at age 2. He was enthroned in Lhasa in 1940. In November 1950, he was obliged to assume full political power following the Chinese invasion of Tibet. When in 1959 a Tibetan national uprising spread to Lhasa and was on the verge of being crushed by the People's Liberation Army, he managed to escape into exile and sought political asylum in India—an event that captured the imagination of the world. Half a century later, he still resides in Dharamsala, the seat of the Tibetan government-in-exile. Awarded the Nobel Peace Prize in 1989, the Dalai Lama is widely regarded as an outstanding international statesman of the Gandhian persuasion—a spiritual mentor of unexcelled moral stature.

Universal Responsibility and the Climate Emergency

It is difficult to fully comprehend great environmental changes like global warming. We know that carbon dioxide levels are rising dangerously in the atmosphere leading to unprecedented increases in the average temperature of the planet. The Earth's great stores of ice—the Arctic, the Antarctic, and the Tibetan plateau—have begun to melt. Devastating sea level rises and severe water shortages could result this century. Human activity everywhere is hastening to destroy key elements of the natural ecosystems all living beings depend on. These threatening developments are drastic and shocking. It is hard to imagine all this actually happening in our lifetime, and in the lives of our children. We must deal with the prospect of global suffering and environmental degradation unlike anything in human history.

If we can begin to act with genuine compassion for all, we still have a window of opportunity to protect each other and our natural environment. This will be far easier than having to adapt to the severe and unimaginable environmental conditions projected for a "hothouse" climate. On close examination, the human mind, the human heart, and the human environment are inseparably linked together. We must recognize we have brought about a climate emergency in order to generate the understanding and higher purpose we need now. On this basis we can create a viable future—sustainable, lasting, peaceful co-existence.

Ignorance of interdependence has harmed not only the natural environment, but human society as well. We have misplaced much of our energy in self-centered material consumption, neglecting to foster the most basic human needs of love, kindness, and cooperation. This is very sad. We have to consider what we human beings really are. We are not machine-made objects. It is a mistake to seek fulfilment solely in external "development."

This blue planet of ours is a delightful habitat. Its life is our life; its future is our future. The Earth, indeed, acts like a mother to us all. Like children, we are dependent on her. In the face of such global problems as the climate emergency, individual organizations and single nations are helpless. Unless we all work together, no solution can be found. Our Mother Earth is now teaching us a critical evolutionary lesson—a lesson in universal responsibility. On it depends the survival of millions of species, even our own.

The destruction of nature and natural resources results from ignorance, greed, and lack of respect for the Earth's living things. Future human generations will inherit a vastly degraded planet if destruction of the natural environment continues at the present rate. Our ancestors viewed the Earth as rich and bountiful. They saw nature as inexhaustible. Now we know this is the case only if we care for it. It has become an urgent necessity to ethically re-examine what we have inherited, what we are responsible for, and what we will pass on to coming generations. We ourselves are the pivotal human generation.

Since the industrial revolution, the size of our population and power of our technology have grown to the point where they have a decisive impact on nature. To put it another way, until the latter half of the twentieth century, Mother Earth was able to tolerate our sloppy house habits. However, our environmental recklessness has brought the planet to a stage where she can no longer accept our behavior in silence. The sheer size and frequency of environmental disasters—Atlantic hurricanes, wildfires, desertification, retreat of glaciers and Arctic sea ice—these can be seen as her response to our irresponsible behavior.

The air we breathe, the water we drink, the forests and oceans which sustain millions of different life forms, and the climate that governs our weather systems all transcend national boundaries. It is a sobering thought that every breath we now take contains more carbon dioxide than at any time for the past 650,000 years—long before the advent of the modern human species. No country, no matter how rich and powerful or how poor and weak, can afford to ignore global warming. It is time for the industrially developed nations to take rapid action to greatly reduce energy waste and to replace fossil fuels with renewable sources such as wind, solar, geothermal, and ocean power. Let us adopt a lifestyle that emphasizes contentment, because the cost to the planet and humanity of ever-increasing "standards of living" is simply too great.

Sometimes we entertain an incorrect belief that human beings can "control" nature with the help of technology. The emergency that climate change represents now proves otherwise: we must restore the balance of nature. If we ignore this, we may soon find that all living things on this planet—including human beings—are doomed.

We are part of nature. Ultimately nature will always be more powerful than us, despite all our knowledge, technology, and super-weapons. If the Earth's average temperature increases by two to three degrees centigrade more than the pre-industrial level, we will trigger a hostile climate breakdown. Morally, as beings of higher intelligence, we must care for this world. Its other inhabitants—members of the animal and plant kingdoms—do not have the means to save or protect it. It is our responsibility to undo the serious environmental degradation caused by thoughtless and inappropriate human behavior. We have polluted the world with toxic chemical and nuclear waste, and selfishly consumed many of its resources. Now we stand at the precipice of destroying the very climate that gave rise to human civilization.

Eminent scientists have said that global warming is as dangerous for our future as nuclear war. We have entered the uncharted territory of a global emergency, where "business as usual" cannot continue. We must take the initiative to repair and protect this world, ensuring a safe-climate future for all people and all species.

The Sheltering Tree of Interdependence

A Buddhist Monk's Reflections on Ecological Responsibility

During the course of my extensive travels to countries across the world, rich and poor, east and west, I have seen people revelling in pleasure, and people suffering. The advancement of science and technology seems to have achieved little more than linear, numerical improvement; development often means little more than more mansions in more cities. As a result, the ecological balance—the very basis of our life on Earth—has been greatly affected.

On the other hand, in days gone by, the people of Tibet lived a happy life, untroubled by pollution, in natural conditions. Today, all over the world, including Tibet, ecological degradation is fast overtaking us. I am wholly convinced that, if all of us do not make a concerted effort with a sense of universal responsibility, we will see the progressive breakdown of the fragile ecosystems that support us, resulting in an irreversible and irrevocable degradation of our planet Earth.

The following stanzas, originally composed in November 1989, underline my deep concern, and call upon all concerned people to make continued efforts to reverse and remedy the degradation of our environment.

O Tathagata
Born of the Iksvaku lineage,[a]
Peerless One who, seeing the all-pervasive nature
Of interdependence
Between the environment and sentient beings,
Samsara and Nirvana,
Moving and unmoving,
Teaches the world out of compassion,
Bestow your benevolence on us!

O Saviour
You who are called Avalokiteshvara,
Personifying the body of compassion
Of all buddhas,
We beseech you, enable our spirits to ripen,
Come to fruition, and observe reality
Devoid of illusion.

Our obdurate self-centeredness,
Ingrained in our minds
Since beginningless time,
Contaminates, defiles, and pollutes
The environment
Created by the common karma
Of all sentient beings.

Lakes and ponds have lost
Their clarity, their coolness.
The atmosphere is poisoned.
Nature's celestial canopy in the fiery firmament
Has burst asunder
And sentient beings suffer diseases,
Previously unknown.

Perennial snow-mountains, resplendent in their glory,
Bow down and melt into water.
The majestic oceans lose their ageless equilibrium
And inundate islands.

The dangers of fire, water, and wind are limitless.
Sweltering heat dries up our lush forests.
The world is lashed by unprecedented storms.
The oceans surrender their salt to the elements.

Though people do not lack for wealth,
They cannot afford to breathe clean air.
Rain and streams do not cleanse,
But remain inert and powerless.

Human beings
And countless beings
That inhabit water and land
Reel under the yoke of physical pain
Caused by malevolent diseases.
Their minds are dulled
By sloth, stupor, and ignorance.
The joys of the body and spirit
Are far, far away.

We needlessly pollute
The fair bosom of our Mother Earth,
Rip out her trees to feed our short-sighted greed,
Turning fertile land into sterile desert.

The interdependent nature
Of the external environment
And people's inward nature,
Described in the tantras,
Works on medicine and astronomy,
Has certainly been vindicated
By our present experience.

The Earth is home to living beings;
Equal and impartial to the moving and unmoving.
Thus spoke the Buddha in truthful voice
With the great Earth for witness.

As a noble being recognizes the kindness
Of a sentient mother

And seeks to repay that kindness,
So the Earth, the universal mother
Which nurtures all equally,
Should be regarded with affection and care.

Forsake wastage.
Do not pollute the clean, clear nature
Of the four elements,
Or destroy the well-being of people,
But absorb yourself in actions
That are beneficial to all.

Under a tree the great sage Buddha was born.
Under a tree he overcame passion
And obtained enlightenment.
Under two trees he passed into Nirvana.
Indeed, the Buddha held trees in great esteem.

The place in Amdo, where Manjusri's emanation,
Je Tsongkhapa's body, bloomed forth,
Is marked by a sandalwood tree,
Bearing a hundred thousand images of the Buddha.

Is it not well known
That some transcendental deities,
Eminent local deities and spirits,
Make their abodes in trees?

Flourishing trees clean the wind,
And help us breathe the sustaining air of life.

They please the eye and soothe the mind.
Their shade makes a welcome resting place.

In the Vinaya, the Buddha taught monks
To care for tender trees.
From this, we learn the virtue
Of planting and nurturing trees.

The Buddha forbade monks to cut
Or cause others to cut living plants,
To destroy seeds or defile fresh green grass.
Should not this inspire us
To love and protect our environment?

They say, in the celestial realms,
The trees emanate
The Buddha's blessings
And echo the sound
Of basic Buddhist doctrines,
Like impermanence.

It is trees that bring rain,
Trees that hold the essence of the soil.
Kalpataru, the wish-fulfilling tree,[b]
Manifests on Earth
To serve all purposes.

In times gone by,
Our ancestors ate the fruits of trees,
Wore their leaves,

Discovered fire by rubbing sticks,
And sheltered amid the foliage of trees
When they encountered danger.

Even in this age of science and technology,
Trees provide us shelter,
The chairs we sit on,
The beds we lie on.
When the heart is ablaze
With the fire of anger,
Fuelled by conflict,
Trees bring refreshing, welcome coolness.

In the tree lie the roots
Of all life on Earth.
When it vanishes,
The land exemplified by the name
Of the rose-apple tree[C]
Will remain no more than a dreary, desolate desert.

Nothing is dearer to the living than life.
Recognizing this, in the Vinaya rules,
The Buddha lays down prohibitions,
Such as the use of water containing living creatures.

In the remoteness of the Himalayas,
In days gone by, the land of Tibet
Observed a ban on hunting, on fishing,
And, during designated periods, even on building.
These traditions are noble,

For they preserve and cherish
The lives of humble, helpless, defenseless creatures.

Playing with the lives of beings,
Without sensitivity or hesitation,
Such as the pursuits of hunting or fishing for sport
Is an act of heedless, needless violence,
A violation of the solemn rights
Of all living beings.

Being attentive to the nature
Of interdependence of all creatures,
Both animate and inanimate,
We should never slacken in our efforts
To preserve and conserve the harmony of nature.

On a certain day, month, and year,
We should observe the ceremony
Of tree planting.
Thus, we will fulfill our responsibilities,
And serve our fellow beings,
Which not only brings us happiness,
But benefits all.

May the force of observing that which is right,
And abstinence from wrong practices and evil deeds,
Nourish and augment the prosperity of the world!
May it invigorate living beings and help them flourish!
May sylvan joy and pristine happiness
Ever increase, ever spread and encompass all that is!

Part II
Global Warming Science: A Buddhist Approach

John Stanley

Climate, Science, and Buddhism

We live at a momentous time. Powerful hostile factors are destabilizing the biosphere—the great global ecological system that integrates all living species and their environment. Our climate system has provided benign conditions for civilization to develop over the last ten thousand years. Now it is entering a critical phase, because humans have released vast amounts of carbon gas through the burning of fossil fuels, and through rampant deforestation. Urgent, decisive corrective action is essential. The risk of inaction is that we could now trigger a runaway global warming process, making most other species on Earth—and even ourselves—extinct. This does not have to happen. The scientific facts, alternative energy technologies, mass communication, and international bodies needed to avoid climate breakdown are in our hands.

It is a fundamental aim of this book to bring Buddhist wisdom and science together in order to arrive at a comprehensive understanding of the climate crisis. The Dalai Lama states:

> In one sense the methods of science and Buddhism are different: scientific investigation proceeds by experiment, using instruments that analyze external phenomena, whereas contemplative investigation proceeds by the development of refined attention, which is then used in the introspective

> examination of inner experience. But both share a strong empirical basis: if science shows something to exist or to be non-existent, then we must acknowledge that as a fact. If a hypothesis is tested and found to be true, we must accept it. Likewise, Buddhism must accept the facts—whether found by science or found by contemplative insights. If, when we investigate something, we find there is reason and proof for it, we must acknowledge that as reality—even if it is in contradiction with a literal scriptural explanation that has held sway for many centuries or with a deeply held opinion or view. So one fundamental attitude shared by Buddhism and science is the commitment to keep searching for reality by empirical means and to be willing to disregard accepted or long-held positions if our search finds the truth is different.[1]

In this spirit, there follows a concise, up-to-date summary of the scientific facts of global warming—from a Buddhist perspective.

Our Own Geological Epoch: The Anthropocene

In order that countless diverse machines might be
brought into service,
There is unlimited excavation of mines, and through
these actions
The abodes of celestial, aquatic, and terrestrial spirits
are imbalanced.
Grant your blessings therefore that afflictions
associated with the elements might be assuaged!
The air is being polluted by billowing clouds of
smoke from countless factories,
And through this primary cause,
The whole world trembles due to unprecedented
diseases.
Grant your blessings that it may be protected from
such states of misery!

—*Sakya Trizin Rinpoche*[1]

Our modern human species is about 150,000 years old. The Earth is 4.5 billion years old and the simplest life forms about 3 billion years old. Life will outlast even an extreme global warming event—as has happened several times in the planet's geological history.[2] This is not true, however, for our present, awe-inspiring biodiversity. This is why UN scientific experts are calling for urgent protection of the planetary life-support systems for humans and most other living species.[3,4]

Regional climates on the Earth vary greatly from the equator to the poles, because the parallel rays of the Sun fall unevenly across the curve of the planet's surface. "Global average climate" is the average of all those climatic regions. It has shown little variation during the last 10,000 years—the Holocene geological epoch that gave rise to human civilization. Before the Holocene, over a hundred thousand years of extreme climate variation allowed only for a hunter-gatherer lifestyle. A stable climate permitted the development of agriculture—and with it, city life, culture, and civilization. Buddhism itself is a product of this stable climatic period.

Stable climate results from a self-regulating atmospheric system. It depends on particular concentrations of greenhouse gases (principally carbon dioxide and methane) in order to partially block radiation of solar energy into space. Prior to the Industrial Revolution of the early nineteenth century, the concentration of carbon dioxide in the Holocene atmosphere was stable at 280 parts per million (ppm).

About two hundred years ago, the Earth entered a new geological epoch. A new industrial revolution powered by carbon fuels began with James Watt's invention of the steam engine (1784). Human economic activity came to dominate the evolutionary path of the planet. For this reason, the Nobel Prize-winning scientist Paul Crutzen named this new epoch the *Anthropocene*[5] from the Greek *anthropos* for man. Oil, coal, and gas are all that remains of algae, plants, and trees that lived by drawing carbon dioxide from the ancient atmosphere hundreds of millions of years ago. When we burn them we release fossil carbon that has been out of circulation for eons. Biologist Tim Flannery appropriately remarked:

> Digging up the dead in this way is a particularly bad thing for the living to do.[6]

The carbon dioxide level in the atmosphere is now 387ppm, the highest for at least 650,000 years. It is rising at an increasing rate, amplifying the greenhouse effect, and so far has increased global average temperature by 0.8°C (1.5°F) over pre-industrial norms.[7] That is enough to have initiated momentous change across all the Earth's climatic regions. The polar ice caps and high mountain glaciers have begun to melt. Extreme weather events (hurricanes and typhoons, floods, heat waves, and droughts) have quadrupled in frequency since the 1950s.

One may well ask why the public is so generally unaware of the global warming threat. Why is there such weak political response? "Frames" are mental structures that allow us to understand or create what we take to be reality. Frames can inform genuine communication, or deliberately confuse collective understanding. For example, the terms climate change, global warming, climate chaos, climate destruction, or climate emergency have different influences on the way we understand this process and its causes. Our mainstream media are contaminated by special interest PR campaigns that reinforce the delusion that fossil fuels remain safe and sustainable. Avoidance of scientific fact combines readily with force of habit, in a world where 80 percent of energy supply comes from oil, coal, and gas. Utilities tied to fossil fuels control electricity generation and distribution. The annual turnover of the oil industry is about three trillion dollars. Nevertheless, big oil corporations still receive annual taxpayer subsidies of hundreds of billions of dollars. Their political influence is backed by sophisticated PR and advertising—in the U.S., at a cost of hundreds of millions of dollars a year.

Coal, oil, and gas rapidly brought about the Anthropocene epoch, and could end it just as quickly. Our massive, increasing global carbon dioxide emissions are already over 20 percent above those of the year 2000. Fossil fuels are pushing the biosphere into climate breakdown. At the same time, they are running out. But they are also generating the largest profits in commercial history. As a result, our society has so far failed to redirect itself toward human well-being, the fate of future generations, and the survival of most other species on Earth.

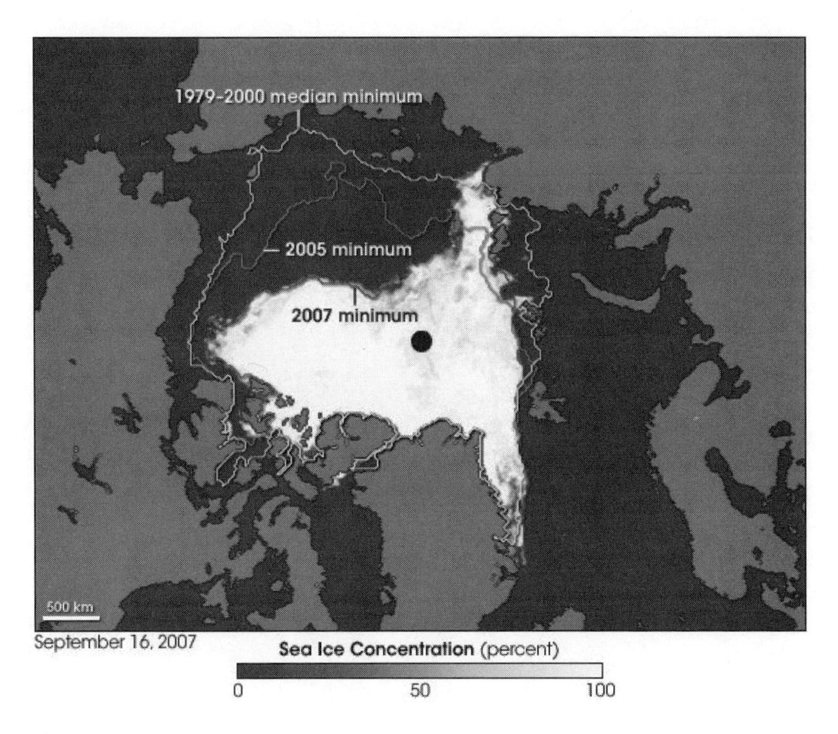

FIGURE 1: This satellite-based image of the Arctic Ocean basin shows the record-breaking 2007 sea ice minimum, in comparison to the previous 2005 record, and the median sea ice extent from 1979–2000. The North Pole is indicated by the black disk. Open ocean (black area) is much darker than sea ice, so it absorbs more heat. The retreating sea ice accelerates warming by "ice-albedo feedback," causing the Arctic to warm twice as fast as the global average. *Courtesy of NASA*

"The celestial order disrupted loosens plague, famine, and war..."

> We are part of nature. Ultimately nature will always be more powerful than us, despite all our knowledge, technology, and super-weapons. If the Earth's average temperature increases by two to three degrees centigrade more than the pre-industrial level, we will trigger a hostile climate breakdown.
>
> —*Dalai Lama XIV*[1]

Today the world economy is seven times as large as it was in 1950. Economic growth and increasing fossil fuel combustion have become inseparable. Within half a century, the Earth's atmosphere has been radically changed by mass transfer of fossil carbon from deep below the planet's surface.

Electricity power stations account for 25 percent of carbon emissions, as does deforestation. Surface transport systems and industry account for 15 percent and 17 percent. Of all the carbon dioxide emitted into the atmosphere, only half of it stays there. The other half goes into carbon sinks—the oceans and land biosphere. The Dalai Lama has stated:

> Scientific predictions of environmental change are difficult for ordinary human beings to comprehend fully. We hear about hot temperatures and rising sea levels, vast population

growth, depletion of resources, and extinction of species. Human activity everywhere is hastening to destroy elements of the natural ecosystems all living beings depend on.[2]

The 2007 report by the Intergovernmental Panel on Climate Change (IPCC) was the collaborative work of 2,000 expert scientists from 100 countries and is a very conservative scientific document. It predicts the following consequences of "business-as-usual" if carbon gas emissions are not radically and systematically reduced. The frequency of major storms and floods will increase dramatically. Sea levels could inundate our coastlines and coastal cities within this century. The oceans will become increasingly acidic, leading to destruction of coral reefs and marine life. Deadly heat waves will become prevalent. Snow will disappear from all but the highest mountains. Agriculture will collapse widely and deserts will expand. Hundreds of millions of people will suffer water shortage and famine.[3] These scientific predictions strikingly resemble the portents described in visionary texts of the Nyingma school of Tibetan Buddhism, such as the *Legend of the Great Stupa*:

> The celestial order, disrupted, loosens plague, famine and war to terrorize terrestrial life... No rain falls in season, but out of season; the valleys are flooded. Famine and hail govern many unproductive years. Diseases, horrible epidemics, and plagues spread like wildfire, striking men and cattle... fire, storms and tornadoes destroy temples, stupas, and cities in an instant.[4]

In actual fact, the Fourth IPCC Report[3] has underestimated the speed of climate breakdown, most obviously in the Arctic. For about 15 million years, year-round Arctic sea ice (an area the size of Australia) and the giant Greenland Ice Sheet have been key elements of the global climate system. In summer 2007, an unprecedented meltdown of Arctic sea ice took place—at least half a century ahead of the IPCC's prediction (Figure 1: p. 41). Sea ice does not increase sea levels when it melts. However, the reflective cooling ("albedo") effect of white ice is replaced by dark, heat-absorbing seawater. Complete loss of Arctic sea ice in summer is now inevitable. Its loss will cause a large local warming in the Arctic region of around 5°C (9°F) and a small but critical global warming.[5]

The land-based Greenland Ice Sheet (GIS) is 2,000 km long, 1,000 km wide and up to 2 km deep. Rising ocean temperatures will move it further toward its critical-melt threshold, a regional warming of 2.7°C (5°F). If rapid, nonlinear disintegration of the GIS occurs, as is possible if the world warms by 2–3°C, sea levels around the world would be increased by one or more meters this century. The oceans would swallow world coastlines and thousands of square miles of productive farmland—from the Mekong Delta to East Anglia. All major port cities of the world would be lost to the sea. New York, Calcutta, Shanghai, Mumbai, Tokyo, Hong Kong, Bangkok, Singapore, Rangoon, Dhaka, and London are among those directly at risk of this catastrophe.[6]

What is now happening in the Arctic has the potential to destabilize the entire global climate system. In September 2008, satellite pictures showed that both north-west and north-east shipping passages had opened up through the perennial sea ice, making it possible for the first time to sail right round the Arctic ice cap. Vast amounts of the very potent greenhouse gas methane could be released as permafrost melts in Siberia and Canada. This would constitute another major climate tipping point. In late 2008, a research ship observed areas of foaming sea off the Siberian coast—"chimneys" where sub-sea permafrost had presumably melted, allowing methane gas to rise from underground deposits formed before the last ice age.

In summary, if we let a summer ice-free state in the Arctic Ocean become permanent, this would kick the climate system into runaway warming. The question then becomes, how can we reverse Arctic warming?[5]

Climate Breakdown at the Third Pole: Tibet

The poison of global warming due to the harnessing
of machines in all places and times,
Is causing the existing snow mountains to melt,
And the oceans will consequently bring the world
within reach of the eon's end.
Grant your blessings that it may be protected from
these conditions!

—*Sakya Trizin Rinpoche*[1]

The Tibetan plateau is the highest and largest in the world. It is some 4 km above sea level, much closer to the stratosphere, and colder than anywhere outside the polar regions. It is guarded to the south by the Himalayas, to the north by the Kunlun, and to the west by the Hindu Kush and Pamir ranges. Its 46,000 glaciers, at an average height of 13,000 ft (3,962 meters) above sea level, are part of the Hindu Kush–Himalaya (HKH) Ice Sheet, the Earth's third largest ice mass. Hence, the plateau has been termed the "Third Pole."[2] Despite its pivotal role in the Earth's climate and water supply, far less is known about it than the Arctic or Antarctic.

The HKH ice sheet is now melting through climate warming, at a rate of about 7 percent a year. Aerial surveys of Tibet over the last thirty years show that snow lines are rising, wetlands are shrinking, and desertification is increasing. In the last half-century, over 80 percent of HKH glaciers have retreated. In the past decade, 10 percent of Tibet's permafrost is thought to have degraded. These changes result from a rise in temperature of 0.3°C (2.5°F) per decade on the plateau—approximately three times the global warming rate.[2] Many scientists conclude that current warming trends and political inertia could lead to complete loss of the HKH glaciers by about 2030.[3]

The winter accumulation of snow on high mountains compresses to form ice caps and glaciers, which are frozen freshwater "reservoirs in the sky"[4] that change volume in response to seasonal temperature and snowfall. In summer, they melt slowly, constantly feeding the rivers below. Their stabilizing, regulatory effect on river flows has been a fundamental feature of our world since long before humans practiced agriculture.

One sixth of the human population now depends for its water supply on glaciers and seasonal snowpacks. From India and China to California, food production is critically dependent on their meltwater. Asia's great rivers—the Indus, Sutlej, Brahmaputra, Ganges, Salween, Irrawaddy, Mekong, Yangtze, and Yellow—derive all-season flow from HKH glaciers. The basins of these rivers are home to billions of people from Pakistan to Indochina.[3]

Shrinkage, melting, and retreat of high mountain glaciers all over the world are striking manifestations of global warming (Figure 2a,b: p. 52). Glacial ice is replaced by dangerous mountain lakes, backed up behind broken rock moraines left by the retreating tongues of glaciers. There are currently more than thirty of these unstable lakes on the northern slopes of the Himalayas. In the short term, outburst floods are inevitable. The much more serious long-term consequence is irreversible decline in river flows, with severe loss of drinking and irrigation water.

In western Tibet lies Mount Kailash (Tibetan: Gang Rinpoche), sacred to Buddhists and Hindus, the source region of the Indus, Sutlej, Ganges, and Brahmaputra rivers. The head of the Sakyapa lineage, Sakya Trizin Rinpoche, reports that ordinary pilgrims have written to him, alarmed by the retreat of its glaciers and snow line. To them, this indicates the disturbance of the protective deities of the mountain. Downstream, in arid Pakistan, 60 percent of the population depends on grain irrigated by the Indus River. If global temperatures rise more than 2°C (5.5°F), this great river will run dry by the end of this century.[3]

Situated in Uttarkhand in the Garhwal Himalaya, the 30 km long Gangotri is one of the largest HKH glaciers. It feeds into the basin of the River Ganges, sacred to 800 million Hindus. Gangotri is shrinking twice as fast as it was twenty years ago. Loss of its glacial meltwater would cause critical, permanent water shortage for 500 million people and 37 percent of India's irrigated farmland.

At the southeastern edge of the Tibetan plateau, the Yangtze, Mekong, and Salween rivers run down from Tibet in parallel, through deeply incised valleys separated by high mountain ranges. In China, 23 percent of the population lives in the regions where glacial meltwater provides a major input to the vitally important Yellow River. Chinese scientists report a major developing threat at its source in Amdo (Qinghai) province:

> Climate change is wreaking havoc at the birthplace of China's mother river. The plight of the Yellow river is a grave warning. Millions of people are at risk from climate change

> and the world must act now to reduce carbon dioxide emissions. If we are to avoid catastrophic climate change, there is not a moment to lose.
>
> —*Li Mo Xuan, Chinese Academy of Sciences*

The Yellow River follows a great bend around the flank of the Amnye Machen mountain range and disperses into a vast wetland that acts as a sponge to regulate year-round water supply for northern China. Senior Chinese climatologist Liu Shiyin states that melting glaciers and permafrost loss are drying out and endangering essential grasslands, lakes, and rivers. Damage to the flora, fauna, landscape, and people of these source regions has a domino effect on the river itself. It will have far-reaching impacts on the economy, society, and people of its middle and lower reaches.

Southeast Asia is a global hotspot for black carbon (BC), the second most important cause of atmospheric warming and glacier retreat on the Tibetan plateau.[2] BC emissions originate from coal-fired power stations, diesel engines, and traditional cooking (with wood, dung, and crop residues). BC particulates remain in the atmosphere for a few weeks, as opposed to carbon dioxide which remains in the atmosphere for more than one hundred years. When they precipitate onto snow and ice, they reduce its ability to reflect sunlight. Hence, reducing black carbon emissions would be an effective short-term way to slow down glacier loss.

> Ironically, the two countries building most of the new coal-fired power plants, China and India, are precisely the ones whose food security is most massively threatened by the

carbon emitted from burning coal. It is now in their interest to try and save their mountain glaciers by shifting energy investment from coal-fired power plants into energy efficiency and into wind farms, solar thermal power plants, and geothermal power plants. China, for example, can double its current electrical generating capacity from wind alone.

—*Lester Brown*[4]

The essential challenge is to design national energy policies to save the glaciers upon which Asian civilization depends. At the time of writing, China and India remain committed to unsustainable policies of economic growth based on carbon fuels. In truth they need to cut BC and carbon gas emissions by 80 percent as soon as possible. Economic policies based around national fossil fuel industries represent an extraordinary danger for billions of their own citizens.

In brief, dependent on strong desire and craving,
This world generated by ordinary past actions
Is beginning to be swiftly transformed into a desert.
Grant your blessings that the negative past actions
which are responsible
Might cease, right where they are!

—*Sakya Trizin Rinpoche*[1]

FIGURE 2A: Meltwater flowing from the Athabasca glacier in the Columbia ice field (Canada). Note the down-cutting of the ice by meltwater flow—when this reaches the underside of the glacier, it lubricates and increases glacial flow. *Courtesy of Robert Rodhe, Global Warming Art Project*

FIGURE 2B: McCarty glacier, Alaska, which retreated 20 km over the twentieth century and is no longer visible in the 2004 photograph. *Courtesy of Robert Rodhe, Global Warming Art Project*

The Road from Denial to Agricultural Collapse

Above all, may individuals recognize that it is their
greatest personal responsibility to implement the
protection of the world's environment!
Grant your blessings that, having recognized this, from
this day on they can practice whatever they preach,
without voicing mere empty words!

—*Dzongsar Khyentse Rinpoche*[1]

Ancient Kyoto, with its classic Japanese Buddhist temples and gardens, seemed an auspicious location to negotiate a treaty to "prevent dangerous anthropogenic interference with the climate system." The well-intentioned Kyoto Protocol of 1997 was fundamentally flawed. It failed to address destructive burning of tropical rainforests, which contributes 25 percent of greenhouse gas (GHG) emissions. It exempted China and India from mandatory reductions, and they developed into major GHG polluters. In 2008, China surpassed America as the world's greatest carbon gas polluter.

Kyoto was rejected by the United States, source of a quarter of world emissions. This reflected the overweening influence of its fossil fuels industry. In 2005, an extreme weather event, Hurricane Katrina, flooded New Orleans. In 2007, the director of the U.S. Hurricane Center, retiring after thirty-eight years of government service, warned that greater disasters were inevitable. The same year, 90 percent of California, Arizona, Nevada, Georgia, Alabama, and Tennessee were suffering the worst drought since the Great Depression. Major wildfires have burned an area larger than the state of Idaho since 2000. Extensive wildfires in California and two hurricanes on the Gulf Coast occurred in 2008. Despite these obvious indications, public understanding of global warming has been comprehensively hampered at the level of both government and media. We can hope that the obstructionist role of the United States toward a global climate protection treaty has finally been reversed by the election of Barack Obama, whose emphatic victory reflected, in part, a popular uprising against corporate control of American government.

Australia is the world's greatest exporter of coal and generates its highest per capita carbon emissions. Its former government refused to ratify the Kyoto treaty even as the country experienced its worst drought in hundreds of years. Over an area the size of France and Germany combined, rivers and reservoirs have run dry. Farmers raise emaciated livestock. Cotton, wine grape, and rice crops have collapsed. The authorities ended further agricultural irrigation to preserve the remaining drinking water. In the 2007 election, Australia rejected the administration that had presided over this ecological disaster.

Australia exemplifies the choice the world will have to make between viable agriculture and the carbon fuels economy. This choice, if properly articulated for the public, could close the fossil fuel industry. Failure to choose swiftly will deny us any real choice, since there is a clear danger of triggering runaway warming. There continues to be no substantive public discussion of the biological fact that most of our crop plants require narrow, specific ranges of temperature and soil moisture. The IPCC predicts that rain-dependent agriculture could suffer a 50 percent yield reduction by 2020.[2] Along with precipitous drops in water tables and the effect of global warming on high mountain glaciers, there is a distinct threat of ongoing famine. This wholly outweighs the so-called difficulty of changing to renewable energy and protecting the remaining rainforests.

The Peak Oil Factor

Oil is the most energy-rich and versatile fossil fuel—the enabler of the modern world. Its potential for generating conflict has led to it being called the "blood of the Earth."[1] Many petroleum geologists consider the peak of global oil production to be now occurring.[2] An oil reservoir is made of almost solid rock. Oil and gas are trapped under very high pressure in tiny interconnected pores. When a well is drilled, the approximately 300-fold difference between the underground and surface pressures forces the oil up the pipe. As it begins to flow, reservoir pressure drops. In due course, it fails to lift the 10,000 foot column of liquid, leaving about half the reservoir's oil behind in the ground.

A typical oilfield has multiple wells. As more are added, production rises to a peak. Then it falls as pressure drop begins to limit oil recovery. In the North Sea, for example, production at the "Forties" oil field grew for five years. It then declined over the subsequent twenty-five years.[2] This kind of scenario holds true for the oil production of an entire geographic region. Accordingly, the date of peak production for the "lower forty-eight" American states was accurately predicted by Texan oil geologist M.K. Hubbert who established that, from well to oilfield to continent, "Peak Oil" occurs at the halfway point of total production. This is an unwelcome scientific fact for the world's most powerful industries, economists, governments, and media outlets.

After Peak Oil occurs, global oil production will enter a phase of terminal decline. Production rates will turn down, along a bell-shaped curve. Oil will not suddenly run out. Rather, the supply of cheap conventional oil drops, while demand continues. Prices rise dramatically, falling back when price temporarily destroys demand. The overall trend is one of declining supply with ongoing price volatility.

Contemporary transport, agricultural, and industrial systems rely on cheap oil. A post-peak production decline, accompanied by continuous price increases, will trigger economic depression. Timely large-scale investment in alternative energy is very important. Such a strategy has been implemented, for instance, by the Swedish government.[3]

An international oil-depletion protocol has been proposed, with the aim of avoiding resource wars, and facilitating a smoother transition to renewable energy. Every nation would aim to reduce oil consumption by at least the world depletion rate and no country would produce or import oil at above its present depletion rate.[4]

> What Buddhist teachers and practitioners can do now is demonstrate how we can break away from the culture of addiction. This will take some strength, because we too can be quite immersed in consumerism. We will need inner discipline to break from it and choose a lifestyle of minimum needs and maximum contentment.
>
> —*Dzigar Kongtrul Rinpoche*[5]

Scientific Predictions of Ecological Karma

> The destruction of nature and natural resources results from ignorance, greed, and lack of respect for the Earth's living things. Future human generations will inherit a vastly degraded planet if destruction of the natural environment continues at the present rate.
>
> —*Dalai Lama XIV*[1]

Ed Ayres has identified four "megaphenomena"—revolutionary changes sweeping the world and transforming everything.[2] Prior to the Industrial Revolution, these phenomena were stable for 100,000 years. Within the last few decades they have "spiked," creating an unprecedented danger for all life on Earth. They are the spikes in carbon gas emissions, consumption, human population, and species extinction.

These four megaphenomena are interdependent and mutually reinforcing. Fossil fuels directly produced the carbon gas spike. Using these fuels increased agricultural productivity and permitted industrial developments that gave rise to spikes in both population and consumption, which then further augmented the spike of carbon gas emissions. These three spikes are synergistic causes of global warming and habitat destruction, ultimately producing the last spike: species extinction.

A number of scientific reports indicate that humans have exceeded the carrying capacity of the Earth by about 25 percent, which means that "biosphere services" have begun to decline.[3] Fossil fuels have permitted levels of resource extraction and waste that mask this ecological degradation. For example, the loss of 30 percent of arable land and soil fertility since 1950 is masked by artificial fertilizers made from natural gas. Although 90 percent of all large fish species have been driven toward extinction, fossil fuels allow fleets to fish ever farther from home. Depleted water tables in India and China are masked by water pumped from tens of millions of deeper wells. As we have already seen with oil, the "one-time inheritance" of carbon fuels is beginning to decline. Within a matter of decades, the extent of ecosystems destruction will be unmasked.

Ecosystem destruction is particularly evident in the tragic burning of great tropical rainforests, the cause of 25 percent of all carbon gas emissions. The Amazon rainforest covered 60 percent of Brazil and 20 percent of that has already been lost. It hosts a quarter of the world's species, accounts for 15 percent of terrestrial photosynthesis and drives global atmospheric circulation. If its total biomass store of 120 billion tons of carbon were released by burning, runaway global warming would be inevitable.[4] The UN Climate Conference in Bali (2007) concluded that new carbon emissions reduction targets should urgently encompass forests, in order to break the cycle of deforestation. However, in that same year, NASA satellite observations showed that ranchers had set more than 10,000 fires across the Amazon to clear land, leading Hylton Murray Philipson of Rainforest Concern to comment:

> These fires are the suicide note of mankind.[5]

The IPCC reports have presented climate change as a smooth, linear transition, but the climate system has rapidly switched states in the Earth's geological history. The juncture where such a state transition is triggered is called a tipping point.[6] Fifty-five million years ago, there was a runaway global warming event called the *Palaeocene-Eocene Thermal Maximum*. It is our closest analogue to current human impacts on the Earth's climate system. The amount of fossil carbon released was similar to our fossil fuel reserves, and came from frozen methane under the ocean. Temperatures rose 5°C (9°F) at the equator and up to 10°C (18°F) at the poles. The ocean acidified and there was very large scale biological extinction. The biosphere took about 100,000 years to recover.

One of the world's leading climatologists, Dr. James Hansen of NASA, states that unless real progress occurs by 2015 to reverse the flow of carbon emissions, the climate system will be propelled past a tipping point.[7] Feedback loops and tipping points whipsawed the early Earth between climatic extremes. Global warming of 2–3°C would lead to a planet with no Arctic sea ice, sea-level rises of several meters, and worldwide super-droughts. Melting of frozen tundra and the destruction of the Amazon rainforest would be likely then to release overwhelming quantities of carbon gases. No conceivable human strategy could reverse runaway global warming once major tipping points are activated. Hansen sums up the critical nature of the climate emergency as follows:

> We have to be smart enough to understand what is happening early on...
>
> What is at stake? Warming so far... seems almost innocuous, being less than day-to-day weather fluctuations. But

more warming is already "in the pipeline," delayed only by the great inertia of the world ocean. And climate is nearing dangerous tipping points. Elements of a "perfect storm," a global cataclysm, are assembled.

The safe level of atmospheric carbon dioxide is no more than 350 ppm. Carbon dioxide is already at 387 ppm and rising 2 ppm per year. The oft-stated goal to keep global warming less than 2°C (3.5°F) is a recipe for global disaster, not salvation. We must draw down atmospheric carbon dioxide to preserve the planet we know. A level of no more than 350 ppm is still feasible, with the help of reforestation and improved agricultural practices, but just barely—time is running out.[8]

The Sixth Great Extinction

A sense of genetic unity, kinship, and deep history are among the values that bond us to the living environment. They are survival mechanisms for ourselves and our species. To conserve biological diversity is an investment in immortality.

—*Edward Wilson*[1]

Major extinction events in Earth history occurred 440, 365, 245, and 208 million years ago. Then the last of the dinosaurs vanished 65 million years ago, probably when a giant meteorite crashed onto the planet, ending the era of reptiles and beginning the era of mammals. This was the Cretaceous-Tertiary event, the fifth great biological extinction of geological history.

Now a sixth spasm has begun, this one a result of human activity. Although not ushered in by cosmic violence, it is potentially as hellish as the earlier cataclysms. If left unabated, it could be the primary cause of extinction of a quarter of the species of plants and animals on the land by mid-century.[1]

The United Nations Environment Program identifies as endangered 1 in 4 mammalian species, 1 in 8 bird species, 1 in 3 amphibian species, and 70 percent of plant species. The global species extinction rate now exceeds the global species birth rate by a factor of 100—soon to be 1,000. Global warming is being accompanied by loss of habitat to deforestation, agriculture, and urbanization. Together these factors could drive half of all species on Earth to extinction by 2100, which would constitute an Anthropocene mass extinction event. From the Buddhist point of view, this unnecessary destruction of our fellow species is an unthinkable loss.

> We must remember, however, that if we act now, it lies within our power to save two species for every one that is currently doomed. If we carry on with business-as-usual, in all likelihood three out of every five species will not be with us at the dawn of the next century.
>
> —*Tim Flannery*[2]

The Dalai Lama has stated:

> We are facing the most massive wave of extinction in 65 million years. This fact is profoundly frightening. It must open our minds to the immense proportions of the crisis we face.[3]

The continuous expansion of economic growth, consumerism, and waste cannot possibly be sustained, because fossil fuels and key industrial resources are running out. The biosphere, our planetary life-support system, can no longer support our levels of exploitation. To irreversibly damage or diminish it is biological insanity. What is it that appears to drive—or manipulate—the human species toward self-destruction? As Wilson states:

> The philosophy of exceptionalism, which supposes that the special status on Earth of humanity lifts us above the laws of Nature, takes one of two forms. The first is secular: don't change course now, human genius will provide. The second is religious: don't change course now, we are in the hands of God, or the gods, Earth's karma, whatever... Every species is a masterpiece of evolution, exquisitely well adapted to the niches of the natural environment in which it occurs. The surviving species around us are thousands to millions of years old. Their genes, having been tested each generation in the crucible of natural selection, are codes written by countless episodes of life and death. Their careless erasure is a tragedy that will haunt human memory forever.[1]

The majority of humans would find it very disturbing to contemplate the degradation of nature and the extinction of innumerable species on this planet. We have an instinctual feeling of kinship and love toward nature and other species that has been termed *biophilia*.[1] Our kinship has been abundantly confirmed through DNA sequencing of the genomes of modern humans, Neanderthal man, and diverse species of bacteria, fungi, plants, and animals. DNA sequencing confirms that chimpanzee and human genomes have many conserved features and are 98.4 percent identical. The chromosomes of humans and other primate species are mosaics of each other. Indeed, all life forms on Earth have a common ancestry, and this shared evolutionary history is encoded in their DNA. The relationship between all life-forms, which is described by the Buddhist theory of interdependence, is now no less than a fully established scientific fact.

Although professional biologists share a unanimous view that climate warming and habitat destruction are creating a mass extinction spike at the present time, global society does not yet acknowledge its existence and implications. Buddhists in particular should be aware that there will not be a second chance to save the Earth's great ecological habitats and systems. The time is now.

> It is the work of all the ages, and it is our work today, more than ever. It is the work that allows one to live, joyously, while in a profound state of grief.
>
> —*Robert Jensen*[4]

> We should see and hear these teachings. We should see and hear the voice of these mountains and rivers, and of the endangered and extinct species. We should see and hear the voice of the atom, the homeless, the children; the voice of the teachings of countless generations past, present, and future. If you try to see with the eye and hear with the ear, you'll never get it. Only if you see with the ear and hear with the eye will you truly be able to see "it" clearly.
>
> —*John Daido Loori*[5]

What Makes Us Do It?

And it bears the fruit of Deceit,
Ruddy and sweet to eat;
And the Raven his nest has made
In its thickest shade.

The Gods of the earth and sea
Sought thro' Nature to find this Tree;
But their search was all in vain:
There grows one in the Human Brain.

—*William Blake*[1]

Humans are one among many social animal species on Earth, and our characteristic social behavior clearly evolved from a primate template. Around 40,000 years ago, a flowering of consciousness among Cro-Magnon people produced the first specialized tools, representational art, musical instruments, and referential language. Notwithstanding this "great leap forward,"[2] a lot of human behavior is problematical and increasingly maladaptive. Our species' track record of self-centeredness, aggression, deception, habitat destruction, and extermination of other life forms is a match for its art, knowledge, science, and technology.

Evolutionary psychology has established the powerful influence of instinct and genetics on human behavior. It also shows that we are far from automatons driven by "genetic determinism." A key human evolutionary adaptation is our ability to vary responses according to circumstances, to learn from experience, and to recognize and exploit opportunities as they arise. Humans are certainly capable of characteristic aggressive behavior, but this trait, like others, can be modified—hyper-aggressive Mongol tribes became a non-violent Buddhist society within a few generations. The capacity for rapid learning and transformation is, more than ever, essential for the survival of our species.

Altruism and cooperation are also instinctual human behaviors.[3] Our closest genetic relatives, the two chimpanzee species, hunt and gather cooperatively, share food, display empathy, practice adoption, and console each other. In humans, reciprocity is a powerful cooperative instinct that makes our complex social systems possible.

> The evidence that human society is riddled with reciprocal obligations is great and growing greater all the time. Like language and opposable thumbs, reciprocity might be one of those things that we have evolved for our own use, but that few other animals have found the use or the mental capacity for.[4]

The "mirror neuron" system in the primate brain creates virtual understanding of others' actions or intentions, providing a basis for imitation, language acquisition, and empathy. From a biological viewpoint, the function of ethical behavior is to protect the human species and its genetic material. We experience deep emotion around this fundamental instinct and transpersonal goal.

In the eighteenth century, Descartes sought to "prove the existence of God" by emphasizing the gulf between animals and humans. His logic, a powerful influence on science, technology, and industrial culture, legitimized mistreatment of the rest of nature and an inflated view of ourselves. Human language, culture, religion, art, and science are truly extraordinary. They do not, however, exist in a vacuum, but evolved from biological and psychological traits we share with other primates.[3]

> We are imprisoned in our small selves, thinking only of having some comfortable conditions for this small self, while we destroy our large self. If we want to change the situation, we must begin by being our true selves. To be our true selves means we have to be the forest, the river, and the ozone layer.
>
> —*Thich Nhat Hanh*[5]

Cartesian assumptions have misplaced us in "splendid" isolation. The future of the human species depends on whether we can rapidly awaken to, and protect, our true evolutionary origin, the natural world. We have only a matter of years to acknowledge our interdependence with it.

> The world may pass a tipping point soon, beyond which it will be impossible to avoid massive future impacts on humans and other life on the planet. Who bears legal or moral responsibility? Scientists? Media? Special Interests? Politicians? Today's public? Our children and grandchildren? Who pays?
>
> —*James Hansen, NASA*[6]

Because of past experience many people feel that politics is something dirty. That is an incorrect concept… if you remain removed from politics just to criticize or complain, that is not a wise way… Get into it and try to change things from within. That is the way.

This planet is our only home. We have no alternative refuge. Therefore everyone has the responsibility to care not only for our fellow human beings but also for insects, plants, animals and this very planet.

—*Dalai Lama XIV*[7]

A Safe-Climate Future

> We are not standing at the threshold of dangerous climate change; we passed through that doorway decades ago. Will we take action, at great speed, to rescue ourselves and the other species with whom we share this planet, or will inadequate action condemn the living Earth to catastrophe?
>
> —*Climate Code Red*[1]

A worthwhile goal makes it possible to imagine and investigate how to move forward—from our present endangered state to one of safety. Five aspects of this path have been described:[1]

1. Our goal is a safe-climate future. We have no right to bargain away species or human lives.

2. We are facing rapid global warming impacts. The danger is immediate, not merely in the future.

3. For a safe-climate future we must take action now to stop emissions and cool the Earth.

4. We must plan a large-scale transition to a post-carbon economy and society.

5. Let us recognize a climate and sustainability emergency, because we need to move at a pace far beyond business and politics as usual.

We must accept the challenge to get back to 350 ppm atmospheric carbon dioxide as a target for the safe-climate zone. The climate sensitivity of the planet to doubled atmospheric carbon dioxide (550 ppm) would likely be 6°C, twice the IPCC's estimate. Conventional thinking proposes targets of 450ppm or higher, but this risks creating an ice-free planet with apocalyptic sea-level rises. If we allow greenhouse gas emissions to rise for another decade, we are taking the gravest of risks with our future. Despite current political discourse, if we remain at 450ppm, we will fully activate "carbon cycle feedbacks" like permafrost melting and carbon-sink saturation. These would take us rapidly and irreversibly up to 700 and even 1000 ppm: a wholly catastrophic outcome.

Science and policy cannot be separated on this vital issue. James Hansen therefore made three key points in his open letter to President Barack Obama,[2] which can be summarized as:

1. We must cut off the coal source, and enact a moratorium on all new coal-fired power ("death factories") until carbon-capture and sequestration becomes a reality, 10–15 years hence.

2. We need a comprehensive carbon tax, across all fossil fuels at their source or point of entry. With a 100% dividend to the public, this will be acceptable and will generate a carbon price adequate to the task.

3. Environmentalists should recognize that we need to fast-track development of "fourth-generation" nuclear power that uses up our nuclear waste as fuel. As a back-up to renewable energy this would be far safer for the planet than coal.

The energy balance of the planet must be restored if we are to avoid dangerous feedbacks and tipping points in the climate system. While this is certainly a difficult problem, it is nonetheless a doable one—so long as we phase out coal. To phase out coal is 80 percent of the solution. Herein lies a clear social implication. If politicians and governments reflexively support the short-term financial interests of coal-mining and coal-burning corporations, rather than taking the action specified by our best scientists as essential for the future of civilization, we must call them to account—through an effective, sustained campaign of citizen action.

Part III
Asian Buddhist Perspectives

Gyurme Dorje

Preface: The Meaning of Aspirational Prayer

The Pali term *panidhi* and the Sanskrit terms *pranidhi* and *pranidhana*, which may be rendered in translation as "earnest wish" or "aspirational prayer," correspond to the Tibetan *monlam* (*smon lam*).[1] This refers to the mental resolve to attain enlightenment and bring an end to the sufferings of all sentient beings, generally undertaken in the presence of a former buddha—actual or visualised. The purpose is to achieve this noblest of all goals, along with lesser objectives that may be in conformity with it. As such, aspirational prayers offered in the presence of a buddha are seeds that will give rise to correspon-ding fruits—some of which may even be the subject of prophetic declarations, given by the buddha in question. Implicit here is the understanding that aspirations will achieve their corresponding results because they are based on the accumulation of past merits, and are not causally unrelated to the fruits to which they aspire.

The entire Buddhist tradition—Theravada and Mahayana—is well acquainted with the four fundamental aspirational prayers, which are designed to develop loving-kindness, compassion, empathetic joy, and equanimity in respect of all sentient beings and the environment.[2] Within the Mahayana, in particular, the cultivation of an altruistic enlightened attitude (Skt. *bodhicitta*, Tib. *byang chub sems*) is considered an essential prerequisite for progress on the path to buddhahood.[3] An important distinction is made between the aspirational or causal phase in which the altruistic enlightened attitude is mentally determined or resolved, and the fruitional phase in which it is actually implemented. In this context, the term aspiration refers to the former. However, there are other Mahayana texts which suggest that aspiration refers to both the cause and the result of cultivating the altruistic enlightened attitude, differentiating between those lower aspirations that seek positive rebirth within higher realms of cyclic existence, those higher aspirations that are dedicated to the well-being of all beings, and those that are intent on purifying buddha-fields through the cultivation of great compassion.

An elaborate description of the bodhisattva's aspiration to attain enlightenment for the sake of all beings is to be found in the *Sutra of the Ten Bodhisattva Levels*, where the "transcendental perfection of aspiration" is described as an attribute associated with the eighth bodhisattva level: the Immovable Level from which regression is considered impossible. Here, an aspiration is made never to be separated from the enlightened mind, wherever one happens to be reborn, and never to cease from pursuing perfect conduct on behalf of all sentient beings.

The same text also enumerates the following ten aspirations that an eighth-level bodhisattva should make: to facilitate offerings in respect of all the buddhas without exception; to maintain the monastic discipline and preserve the Buddhist teachings; to perceive the significance of all the incidents in the early career of a buddha; to realize the mind of enlightenment, while acquiring all the transcendental perfections and purifying the ten bodhisattva levels; to mature all six classes of sentient beings and establish them in buddhahood; to perceive the whole universe; to purify all buddha-fields; to enter the Mahayana and foster a common purpose in all bodhisattvas; to ensure that all actions of body, speech, and mind are fruitful and successful; and to teach Buddhism, having attained enlightenment.[4] More briefly stated, the aspirant bodhisattva should seek nothing less than enlightenment and the well-being and liberation of the entire world.

Within the Nyingma tradition of Tibetan Buddhism, the *Tantra of the Great Natural Arising of Awareness* (*Rig pa rang shar gyi rgyud*) refers to the fruitional aspect of perfect aspiration in the following verse:

> The ends of the perfection of aspiration
> Are subsumed in the absence of hope and fear
> With respect to all phenomenal appearances![5]

Gyalwang Karmapa XVII, Orgyen Trinley Dorje Rinpoche (b. 1985), is revered as an extraordinary embodiment of compassion, the seventeenth incarnation of Karmapa Dusum Khyenpa—the oldest of all the incarnating lineages of tulkus in Tibetan history. Born into a nomadic family in Lhato, Kham, in 1992 he was enthroned at Tsurphu Monastery near Lhasa as the head of the Karma Kagyu order of Tibetan Buddhism. By the age of 10, he had begun to recognize other major tulkus within this tradition and, following his daring escape to India in 2000, he has resided in the vicinity of the Norbulingka Institute, near Dharamsala.

Gyalwang Karmapa XVII, Orgyen Trinley Dorje

Pure Aspiration, Bodhisattva Activity, and a Safe-Climate Future

For most of human history, people all over the world had a simple lifestyle that made use of natural resources sustainably, and avoided significant damage to the Earth. In recent times, however, our lives and our relationship with the environment have become increasingly complex—and problematical because we now have tremendous power to harm the living world.

The lifestyle of the late twentieth and early twenty-first centuries is making huge demands on the natural environment. We make unprecedented use of resources such as water, wood, and soil, without correctly understanding what the outcomes will be. In particular, we use fossil fuels recklessly, ignoring the fact that they cause ever higher carbon dioxide emissions, and therefore dangerous global warming as a result. We imagine we need all kinds of cleverly-advertised consumer products, without really evaluating whether they are truly important or useful to us. There seems to be no limit to human desire, but there is clearly a limit to how much Mother Earth can sustain our greed.

The Buddha and his original monastic community followed a way of life that was mindful, frugal, and without waste. It did not fall into the extremes of poverty or hoarding, and it manifested the key principle of the Middle Way. Our lifestyle today should be modelled on this principle—neither too hard nor overly indulgent. If something we desire is beneficial and does not harm the environment it could be considered necessary. If that is not the case, let us think twice about whether we want or need it at all. As Ashvagosha said in his *Twenty Verses on the Bodhisattva Vow*:

> For others and also for yourself,
> Do what is useful even if painful,
> And what is both useful and pleasurable,
> Not what gives pleasure but is of no use.

This active decision-making process represents a choice made out of awareness, rather than made blindly. Our actions match our spiritual aspirations.

Our aspiration as Dharma practitioners is to free all living beings from suffering. Wherever there is suffering, we wish to transform it into happiness and equanimity. We understand that our sense of self is misleading. In reality, the self is not independent from the rest of life around us, even the air we breathe. The principle of interdependence shows us that all life is connected, and that our individual actions have consequences in the larger world. This is the karmic relationship between cause and effect. It clearly applies also to global warming, which has been caused by humans extracting fossil fuel reserves laid down hundreds of millions of years ago and burning them to produce heat, mechanical, and electrical energy. By doing this we have released fossil carbon gas into the atmosphere of our planet.

As Dharma practitioners we have a responsibility to reverse negative actions through skillful means. To ensure that there is a healthy future for all life on Earth, we should be in the forefront of efforts to reduce carbon emissions and replace fossil fuels with renewable sources of energy—wind, solar, (appropriate) water, and geothermal power. We should also take a lead in the protection of forests and rainforests. Their destruction contributes greatly to climate breakdown, while their preservation cools the Earth and ensures its biodiversity. Indeed, we should be part of a global effort to plant many more trees and forests.

I grew up in a remote area of Tibet following a centuries-old way of life. People used water, wood, and natural resources with great care and they generated little or no waste. Even as a child, I planted a tree to protect our local spring and asked my father to protect it when I left for my monastic seat at Tsurphu. We had little formal education, but we inherited a deep traditional concern for our environment. Even as children we regarded many mountains, rivers, and some wild animals as sacred, and treated them with respect accordingly. Now scientists tell us that if we do not make fundamental changes in the way we do business as a global society, we stand to lose over half the species on Earth by the middle of this century. Is this not unbearably sad? Can we not do better than stand idly by, when we know this process of mass extinction is taking place in our own lifetime?

Climate breakdown is already impacting all our lives, and without urgent corrective action it will only become more devastating. Here in the Himalayan region, our climate is warming three times faster than the global average rise in temperature. This is having dire consequences for our great glaciers, which are part of the third largest store of ice on Earth—the so-called

"Third Pole." Upon it depend the ecology and way of life of Tibet, together with the water supply and food of billions of people in China, India, and Pakistan.

We humans have already done such immense damage to the environment that it is almost beyond our power to heal it. The challenge is far more complex and extensive than Buddhists can tackle alone. However, we can take a lead, and to do so we must educate and inform ourselves. This is the time when our pure aspirations and our bodhisattva activity must come together. This is the time to ensure a safe-climate future for our planet. This aspiration comes from my heart.

THE BODHI TREE, a revered sacred fig tree *(Ficus religiosa)* at the Mahabodhi Temple in Bodh Gaya, India. It is a direct descendant of the original tree under which the Buddha achieved enlightenment, and is considered the most important of the four sacred places of pilgrimage for Buddhists. After his enlightenment, the Buddha is recorded to have spent a whole week gazing at the tree with gratitude. The Dalai Lama has written (in this volume):

> Under a tree the great sage Buddha was born.
> Under a tree he overcame passion
> And obtained enlightenment.
> Under two trees he passed into Nirvana.
> Indeed, the Buddha held trees in great esteem.

Kyabje Sakya Trizin Rinpoche (b. 1945), Ngawang Kunga, is the forty-first patriarch of the Sakya lineage and a married lama with two sons. Displaying remarkable intelligence and incisive discriminative awareness from an early age, he began teaching and presiding over tantric empowerments from the age of eight. He was enthroned as the throne-holder of Sakya (Sakya Trizin) by Dalai Lama XIV, prior to his escape to India as a refugee at the age of 14. In exile, he re-established the main seat and college of his tradition at Rajpur, India, in 1964. Since then he has continued to revitalize the sutra and tantra traditions of the Sakya order, in particular, and Tibetan Buddhism in general. Traveling extensively around the world, he confers empowerments and gives teachings in the English language, as well as in Tibetan.

Kyabje Sakya Trizin Rinpoche

The Global Ecological Crisis: An Aspirational Prayer

When I pray one-pointedly, with fervent faith and devotion,
To the master Padmakara, to Padmapani, venerable Tara, and other deities
In whom the Three Precious Jewels are all gathered,
I beseech you to direct your enlightened intention compassionately toward us
From the invisible expanse of reality!

All eons of time that are illustrious in lifespan and merits are destroyed
By ill-intentioned thoughts and deeds, and by evil barbarity.
Will you not therefore direct your enlightened intention compassionately
Toward living beings who, lacking positive opportunities,
Commit an enormous mass of degenerate actions, embodying the five degradations?

Due to rapacious greed that covets the world's
resources
Trees and forests are cut down and so forth,
Causing an imbalance of the rain-water element.
May you swiftly and compassionately protect
Living beings who fall into such disastrous
circumstances!

In order that countless diverse machines might be
brought into service
There is unlimited excavation of mines, and through
these actions
The abodes of celestial, aquatic, and terrestrial spirits
are imbalanced.
Grant your blessings therefore that afflictions
associated with the elements might be assuaged!

The air is being polluted by billowing clouds of
smoke from countless factories,
And through this primary cause,
The whole world trembles due to unprecedented
diseases.
Grant your blessings that it may be protected from
such states of misery!

In particular, due to insatiable desires and cravings,
Coarse human behavior pulverizes the physical world
and its organisms,
Giving rise to an imbalance of the four naturally
occurring elements.

Grant your blessings therefore, that the mundane
aggregates
Might be pacified right where they are, without
causing harm!

The poison of global warming due to the harnessing
of machines in all places and times,
Is causing the existing snow mountains to melt,
And the oceans will consequently bring the world
within reach of the eon's end.
Grant your blessings that it may be protected from
these conditions!

Moreover, there are incurable skin diseases that arise
From the breaching of the natural ozone canopy
Which inhibits the intolerable and terrifying
poisonous radiation of the sun.
Grant your blessings that these may be pacified,
remaining behind in name alone!

In brief, dependent on strong desire and craving,
This world generated by ordinary past actions
Is beginning to be swiftly transformed into a desert.
Grant your blessings that the negative past actions
which are responsible
Might cease, right where they are!
Although the entire mass of defects that afflict the
physical world and its living organisms
Has been engendered by the dissonant mental states
associated with past actions,

Comprising all primary and secondary dissonant
mental states,
Even so, through the unfailing power of truth, of the
Three Precious Jewels,
I pray that all the points of this aspirational prayer may
be fulfilled!

This prayer was composed by Ngawang Kunga of the Dolma Palace, throne holder of Sakya, at the request of Dr. John Stanley, conveyed with the great clarity of higher aspiration, with regard to the impending catastrophe that now confronts the environment and living organisms in all parts of the world. May its aspirations be accordingly fulfilled!

MOUNT SHISHAPANGMA: Framed in a circle of fluttering prayer flags, Shishapangma (8,012 meters, 26,282 feet) is the highest mountain entirely within Tibet and guards a principal gateway from Nepal. This view is from Yakrushong La pass (5,150 meters) on the Nyalam-Dingri road. Prayer flags exemplify the Tibetan Buddhist practice of sanctifying the environment through harnessing the spiritual power of sacred sound and positive intention to that of the wind. *Courtesy of Matthieu Ricard*

Kyabje Dudjom Rinpoche, Sangye Pema Zhepa (b. 1990), the reincarnation of Kyabje Dudjom Rinpoche (d. 1987), the first head of the Nyingma tradition in exile, was born in eastern Tibet. His own father, Dola Jigme Tulku, was one of the late Dudjom Rinpoche's sons. He was initially recognized by Khandro Ta-re Lhamo of Nyenlung in Sertar, and by Dzongsar Jamyang Khyentse Rinpoche. Chatral Rinpoche, Mindrol-ing Trichen Rinpoche, and Sakya Trizin Rinpoche, among others, subsequently concurred, and he was enthroned as Dudjom Tulku at Godavari, Nepal, in 1994. His root teacher is the great yogin Kyabje Chatral Rinpoche.

Kyabje Dudjom Rinpoche

The Mandala of the Four Energies in the Kaliyuga

If one is a follower of Mahayana or the Great Perfection, it is excellent to rest in the view of Emptiness or the view of the Great Perfection, where everything is complete and perfect. On the other hand, on the relative level, we are people who have taken birth. The reason we took birth was basic ignorance. So once we are already in the whole machinery of samsara, this world of suffering, where we are born and subject to the environment we live in, then of course we need to pay respect to that, and take care of it.

It is important to investigate what is happening. Scientists have established and continue to provide more detail about the causes and reasons for global warming and ecological degradation. As much as we are able to rid ourselves of all the pollutants concerned, we should actively try to resolve this crisis. We are products of this world and we live in interdependence with it. Thus, it is very important from the relative perspective to be engaged in solving our most important environmental issues.

Many highly realized masters of the past prophesied that events could occur in the Kaliyuga (our present era), such as the melting of great mountain snow caps and glaciers, and other disasters involving the four elements. Peoples' activities in general have changed in ways that have brought about global warming. In the relative dimension, this is indeed the case. The environment has been polluted, underground water is drying up, ancient ice and snow mountains are melting. Human beings are the cause of such things happening.

Future energy sources will have to be environmentally benign—such as wind and solar. If these innovations are applied, and can be enforced, it will help everybody. If it can be established that renewable energy is viable and cost-effective, we should ask why our societies continue to use coal and other dirty energy sources. If there are viable ways of using solar and wind energy, everyone should concentrate on putting them into use. There needs to be more of a push toward renewable energy.

Buddha taught that all causes and outcomes are created through our own mental constructs. If we could control our negative thinking, everything would proceed on a more constructive level. If the cause is good, the effect is that you yourself will benefit. Great teachers have pointed out that the world is not changing, peoples' minds are changing. Since peoples' confused thinking is increasing, it creates all the kinds of counterproductive outcomes that society is experiencing at this time. Buddhism recognizes that the world environment has not changed by itself—it has been changed by human error or greed. With respect to the role of fossil fuels and the energy industry, we need to combine the latest scientific facts and figures with the way of thinking of the Buddha to bring about a good resolution.

It seems that renewable energy is possible, but those forces that seek to perpetuate the use of fossil fuels are still too strong. It would be very difficult to change all these things at once. If we want to climb upstairs, we have to go step by step. If we build a house, first we lay foundations, and that takes time. Scientists and others should work together to progressively establish the benefit of new, harmonious energy sources. We have to make real effort to achieve the benefits of renewable energy. It is probably not possible to change everybody's attitude immediately. Yet I think that cooperative, progressive efforts will lead to better results in the future.

We place our confidence in the three jewels of Buddha, Dharma, and Sangha and there is definitely great positive benefit in performing certain pujas, or ceremonial offerings. Nonetheless, everything depends on the individuals, who have to improve their own way of thinking. Buddha said, "I have shown you the path that leads to enlightenment, but you yourself have to use that path." Let's say we ignore the spiritual path. At the time of death, we experience a confused and disturbed state. Although we appeal for help, even a buddha cannot rescue us, because our own negative karmic deeds have brought about that suffering. So there are these two aspects: individuals have to improve their way of thinking and, without doubt, if we could practice pujas like Drolo and Phurba (wrathful deities known for eliminating obstacles) or Jinseg (a burnt offering) they could benefit the whole world now.

This world evolved on the mandala of the four energies. Accordingly, in the Kaliyuga there is the possibility of destruction by the elements of heat, water, and wind. We are living in a kalpa, or eon, that ultimately will come to an end. Great masters have prophesied it; it is all in our texts. We cannot specify the time-frame. This world's existence has gone on for billions of years, so to pinpoint an end-phase precisely is not possible. Science and Buddhist philosophy are complementary, so using them together, while not harming sentient beings or the living world—this will benefit beings and be really positive. If we can establish the facts correctly, and change the general attitude toward the living world and renewable energy, that could benefit all beings, not only humans. It is a worthwhile, noble cause. I am happy to support it.

A Prayer to Protect the Earth

Homage to the spiritual teacher of sages and gods,[d]
Whose buddha-mind renounced deceitfulness many
eons before,
Who one-pointedly loves living beings with kindness
and compassion,
And who attained the supreme level of accomplish-
ment through the direct path!

O Compassionate Teacher,
You who have manifested the abiding nature of all
that can be known,
Through pristine cognition that perceives without
obscuration or corruption,
Skilled in teaching the disposition of dependently
arisen phenomena—
Turn your enlightened intention toward this very time!

Now when the elements of the physical world and its
inhabitants have degenerated
Due to bad habits born of greed and aversion, with
regard to mundane prosperity,
And through the production of various man-made
chemical substances,
The time has come for you to love us even more with
your compassion!

At this juncture when the world and its inhabitants
near dissolution,
As life forms, snow mountains and continents are
degraded
By the excavation and ravaging of elemental
resources,
By the polluting fumes of some electricity generation
technologies,
And by various contagious diseases, harmful to living
beings,
We pray with fervent devotion
To the Lake-born Lord, sole refuge of this degenerate
age,
To Avalokiteshvara, most sublime being of compassion,
To Tara, Mother of the Conquerors, who protects
from the eight fears,
And to all the ocean-like hosts of the Three Roots
and oath-bound protectors:
Do not renege on your compassionate pledges to
protect us
(Just by calling your faces to mind)

From disharmony, degradation, and the like,
Caused by engaging in negative actions!

Grant your blessings that the world might enjoy the glory of peace and happiness
And that all living beings might swiftly and without obstruction
Accomplish all their aspirations that accord with the sacred teachings, just as they wish!

At the request of Monlam Gyatso (Dr. John Stanley), who is a student of the previous Kyabje Dudjom Rinpoche, I, Sangye Pema Zhepa, who have been given the name Dudjom Tulku, composed this aspirational prayer on March 26, 2007. May virtue prevail!

MOUNT EVEREST (Tibetan: Jomolangma), the highest mountain on Earth at 8,848 meters (29,029 feet) above sea level, is located on the Tibet-Nepal border. In this view from the Tibetan side, the central Rongpu glacier can be seen. Himalayan glaciers are the source of the major river systems of Asia. They are rapidly declining in mass and extent, not only through global warming caused by world carbon dioxide emissions, but also through the loss of their reflective "albedo effect" caused by black carbon emissions in Asia. *Courtesy of Gyurme Dorje*

Kyabje Chatral Rinpoche (b. 1913) is a reclusive Nyingma master known for his great realization, strict discipline, and integrity. Affiliated to Katok Monastery in his native Kham, he received his name (which means "indestructible buddha who has abandoned all mundane activities") from the renowned Khenpo Ngagchung (d. 1941) of Katok, and also became a major lineage holder of the Longchen Nyingthig and Dudjom Tersar traditions. He was appointed as a principal teacher of the Regent of Tibet in 1947, and following his establishment of vital meditation hermitages in India and Nepal, became renowned throughout the Himalayan regions, where, at the present day, he is regarded as one of the greatest living Dzogchen masters. Shunning the distractions of foreign travel and fame for the pursuits of spiritual practice and isolated retreat, he continues to exemplify the simple life of a yogin, even in his mid-nineties.

Kyabje Chatral Rinpoche

A Prayer at a Time of Ecological Crisis

Sugatas and bodhisattvas of the ten directions,
Turn your enlightened intention toward us!
May all sentient beings tormented by this present age
Of the five virulent degenerations,
Know that they possess a treasure that can alleviate
The various portents of decay in the physical world
and its inhabitants
Due to the ripening of their wrong intentions and
actions—
A treasure grounded in the renunciation of harmful
actions
And the cultivation of altruistic actions,
Granting all the spiritual and temporal well-being one
could desire.
This is the supreme wish-fulfilling gem of good heart
Associated with all supreme spiritual practices.

Endowed with this (good heart),
May all beings cultivate love and compassion for one another,
Without hatred, and without fighting or quarrelling.
May they enjoy the glorious resources of happiness—
All they could possibly desire,
And swiftly attain the level of conclusive omniscience!

This aspirational prayer was written by Sangye Dorje on the tenth day of the second lunar month of the Fire Pig year at the insistence of Dungse Kunzang Jigme Namgyal, who presented a "good day" offering scarf. May it be auspicious!

A THREATENED LIFESTYLE: A nomad family in eastern Tibet in their summer pastures. Female yaks *(dri)* calve in early summer, the only season they can graze sufficiently to feed their young. This ancient, ecologically harmonious way of life is gravely threatened by rapid climate warming on the plateau. Some 10 percent of Tibet's permafrost, essential for the viability of the high alpine flora of this bio-geographical area, has already been lost. *Courtesy of Matthieu Ricard*

Khenchen Thrangu Rinpoche (b. 1933), the ninth reincarnate lama of Thrangu Monastery, near Jyekundo in Kham, has re-established many important oral transmissions and republished many rare texts which had been lost during the turmoil of Tibet's occupation and the Cultural Revolution. Exiled in India, he was appointed as the head of Rumtek monastery in Sikkim, the seat of the sixteenth Karmapa, and as the director of its Institute for Higher Buddhist Studies. A pre-eminent Kagyu lineage holder, he is one of the most celebrated living exponents of the Mahamudra teachings. Consequently, in 2000, the Dalai Lama appointed him the personal tutor of Gyalwang Karmapa XVII. It was his composition of an Aspirational Prayer in 2006 that initiated this book.

Khenchen Thrangu Rinpoche

When Snow Mountains Wear Black Hats

(This dialogue took place at a public seminar in Oxford, U.K., in September 2007, and was translated by David Karma Choephel.)

Question: Scientists predict that if global warming continues unabated, half of all currently living species will become extinct within this century. The global environment will be unable to support the human population we know today. How would this affect the field of reincarnation? What are the implications for Buddhism and for sentient beings?

Thrangu Rinpoche: The first of your questions concerns the danger to the Earth as a place where beings can be reincarnated. I do think this is a great danger. If the external world we live in does not thrive, how can that be good for the "internal" beings who live in that world, its inhabitants? In the external world there are clearly problems. For instance, this year when I was in America, I went to Colorado and I was really surprised to see that all the trees were turning brown and withering. In the valleys and on the hillsides we saw that the trees had dried out and were dying. We need to do something to help that situation. Whose responsibility is it? It is the responsibility of all the "wandering beings," the humans, to do that. It cannot be the responsibility of animals. Animals do not have the power to do anything about the situation. Those of us who are humankind have the responsibility to take care of this world we are in. We might say that it is the responsibility of governments. But whether governments will do anything or not is another question. We all individually need to do something about it.

The Buddha taught the Dharma 2,550 years ago. At that time, there was no such danger. At that time India was a wealthy place and a vast, pristine area. Many merchants would go off in ships, across the sea, find jewels and precious materials, and bring them back to sell in India, where they were valuable. It was such a prosperous place that one thousand monks could all go into the city asking for alms, and a single household would feed them all. They did not have these sorts of difficulties. For that reason, in Dharma we do not find classical teachings that directly address global warming.

Q: Great Buddhist masters such as Patrul Rinpoche referred to this historical period as the degenerate Kaliyuga or "age of dregs." What meaning do these terms have in relation to the destruction of our climate by global warming?
TR: They are closely connected. When we call this the degenerate age or the age of dregs, we mean this is a time when sentient beings are not easily satiated. They are not modest in their wishes. So they do a lot of business in order to benefit themselves. They make a lot of pollution to do business and gather wealth. They do not gather that wealth for the benefit of the whole of society, but for their own individual benefit. In doing so, they pollute the

ground, the water, and the air. It creates a problem for the whole world. It is all really due to our greed. "Degenerate age" arises from the strong negative emotions we have. This is something we should think about carefully.

Q: The great Buddhist teacher Padmasambhava (Guru Rinpoche) made reference to a dangerous future "when snow mountains wear black hats." Do you think this may refer to the disappearance of the high mountain snows and glaciers due to global warming?
TR: Yes, this probably is related to global warming. Guru Rinpoche was making his own predictions about the future, and describing what is going to happen.

Q: In Buddhism we have the practice of saving the lives of animals. How does this inform what we, the Buddhist community, should do to help avert the evolutionary catastrophe that may result from (unabated) global warming, one that could destroy half of all living species by 2050?
TR: We protect the lives of sentient beings through life-releases or saving the lives of animals directly. So, if we can protect sentient beings by reversing global warming, this is a really fortunate thing to do. Therefore, we should definitely try to stop or reverse global warming. We need to know what is happening to our world, what scientists have elucidated. When we know this, we can infer what we need to do, and what we actually can do. Understanding how things are and working to change it is one aspect. Another is to pray to the Three Precious Jewels—make supplications and aspirational prayers.

Will making prayers and reciting aspirations directly stop global warming? It is not going to directly stop global warming.

However, it will gradually transform our mind, transform every mind—and we will make progressive efforts to transform the situation. Such activities do actually help the situation. Making prayers and aspirations is by no means pointless. We need to learn about ecology, train ourselves, and come to understand how to stop global warming. On the other hand, we need to recite aspirational prayers. This is how we can progressively help and be of benefit.

Q: What is the role of Buddhist prayer and puja to reverse, in the invisible realm, the potential disasters of global warming?
TR: In general, when we talk about making aspirations and prayers in Buddhism, it looks superficially like mere blind faith. You might think there is no way it could help at all. Yet if we look at it closely, it really does help. This is my experience. For example, I have a well-off student in Malaysia who built a thirty-two-story hotel. No matter how hard he worked, nothing came of it. One partner was less than honest and he simply lost lots of money. He asked me what to do about the problem and I suggested he recite the Verses of the Eight Noble Auspicious Ones daily. He learned it by heart and did so. The hotel ended up a complete failure—but his other property in Australia jumped so much in value that he recouped all the losses in Malaysia. He told me, "Reciting that prayer was really wonderful!" This is just to show that reciting such prayers is not blind faith, but something genuinely beneficial.

Q: Tulku Urgyen Rinpoche observed in his memoirs that Tibetans simply did not want to hear about the impending Chinese invasion of Tibet in 1950, even though there were obvious and abundant signs. Is there a parallel with our response to global warming now? What can Buddhist practitioners do to help the developing situation?

TR: This is much the same phenomenon Tulku Urgyen was describing. Because it is a parallel situation, we need to wake people up about global warming. We should make aspirations and recite prayers. And we need to have a lot of publicity. These two together will lead to powerful, effective action.

Q: The Dalai Lama is "offsetting" carbon emissions from flights for his world teaching tour. Should other teachers follow his example? Could we use a carbon-offset scheme to directly fund monasteries in the East to go "off the grid" with rooftop solar electricity generation?

TR: Decreasing the amount of carbon gas we produce is not something only those spending money should do; it is everybody's everyday responsibility. All of us have to work together to reduce carbon emissions in the world. Meanwhile, if some people give money to offset their carbon emissions, and use it to help out in poor countries, that would be good.

Q: Can Buddhists contribute by "greening the mind": reducing consumption as an example to others?

TR: Buddhist teachings speak a lot about being satisfied and content. This is something universally applicable and helpful. Even those who have wealth need to learn to be content with what they have. If we can share with others the Buddhist teachings on being satisfied and content, we are giving them something truly valuable.

Q: Among Christian leaders, the Russian Orthodox Patriarch considers that global warming is the greatest moral and ethical issue of our time. He has said that all religions should come together at this time, over this issue. Do you agree?
TR: If we can all work together, that will be excellent. As for Dharma or religion, the point is to bring happiness to the world, to free the world of suffering. That is what is really important in Buddhism, and also in other religions, all of which try to bring happiness to this world. We should train in ways to overcome the difficulty and suffering of sentient beings. It is important to make prayers and aspirations, training in this way. It is important to cooperate with other religions, rather than contradicting each other or disputing in any way. We should act together. We should cultivate a gentle way of giving advice.

Q: Could you give a commentary on your own "Aspirational Prayer to Avert Global Warming"?
TR: I myself am someone who has been born in this world. I do not have the power to do something about global warming all by myself, but it is my responsibility to help people. So I have written a short poem of aspiration on the subject. I will share the meaning with you.

It begins with a supplication to the exalted sources of Refuge: the Buddha, Dharma, and Sangha, the Three Jewels, and also to the Lama, Yidam, and Protectors, the Three Roots. In Buddhism in general, the source of our refuge or protection is the Three Jewels. In secret mantra Vajrayana, the Three Roots are the source of refuge. So here is a prayer that in this world we may pacify the terrors of illness in the first instance, famine in the second, and war in the third. May the blessings of the Three Jewels pacify all these.

What is happening today is that there is chaos in the elements. The four different elements of earth, water, fire, and air have now become unbalanced. Due to this imbalance, sometimes there is destruction because of floods and water, sometimes due to wind. Sometimes there is destruction through earthquakes and now there is also destruction happening through global warming. The temperatures are unbalanced—sometimes too hot, sometimes too cold. Because of this, the grand, resplendent snow mountains are melting. Hard, firm, beautiful glaciers are melting. When they melt and disappear, the rivers and lakes will become scarce: parched and dried out. We can actually see this happening. Disturbance in the four elements, snow mountains and glaciers all melting, the water all drying up—what harm do these lead to? Then, all the forests of the ancients and trees of beauty will near their deaths. Beautiful, wonderful forests that you could explore are drying out, and coming to their deaths. There is now the danger that the whole world's reaches will become a great wasteland without water supplies. In the entire world, we will not have anything beautiful or good, nor any way of supporting ourselves. There is a frightful, terrifying danger that this will happen. It is a basic truth that if something bad happens to the environment we live in, the inhabitants that live within it will also suffer great harm.

Following that, there is an auspicious prayer that these difficulties and problems be resolved, that they may be pacified. When they are pacified, may our good fortune and happiness spread all around. May our outer good fortune—happiness, wealth, resources—develop. May our internal qualities develop. And may all beings nurture one another lovingly and kindly, so their joy may fully blossom. May all sentient beings show loving-kindness to each other. May they also take care of and nurture one another. In this way, may all their aims be fulfilled in accord with the Dharma. I wrote the prayer with hopes that it would help people. My aspiration is that it may help many sentient beings.

An Aspirational Prayer to Avert Global Warming

May the blessings of the exalted sources of refuge—
The Buddha, his teachings, and community, the Three Precious Jewels,
And the spiritual teacher, meditational deities, and protectors of the Buddhist teachings, the Three Roots—
Fully pacify the terrors of illness, famine, and war,
Along with chaotic disturbances of the four elements:
The imminent and terrifying danger that the whole world will become a great wasteland,
As temperature imbalance causes the solid glaciers of snow mountain massifs to melt and contract,
Afflicting rivers and lakes, so that primeval forests and beautiful trees near their deaths!

May the sublime endowments of good fortune and spiritual and temporal well-being flourish,
And may all beings nurture one another lovingly and kindly,
So that their joy may fully blossom!
May all their aims be fulfilled, in accordance with the sacred teachings!

At the request of the scientist John Stanley, a student of the Lord of Refuge Dudjom Rinpoche, this aspirational prayer was composed by Tranguwa.

Dzongsar Jamyang Khyentse Rinpoche (b. 1961), the principal incarnation of Jamyang Khyentse Chokyi Lodro (d. 1959), was recognized by Kyabje Dilgo Khyentse Rinpoche, the root teacher among his several illustrious masters. He has established teaching centers in Australia, North America, and the Far East, and revisited his traditional seat at Dzongsar Monastery in Kham, while continuing to supervise new colleges in India and Bhutan. His international organization is called Siddhartha's Intent. He is a noted filmmaker and the author of *What Makes You Not a Buddhist.*

Dzongsar Jamyang Khyentse Rinpoche

A Prayer to Protect the World's Environment

Let us make offerings to the Three Precious Jewels,
the Ocean of Conquerors,
And especially to Guru (Padmasambhava) Nangsi
Zilnon,[e] sublime Avalokiteshvara, and venerable
Cintamanicakra (Tara), who are our allies in this
degenerate age!

Let us seek refuge in them!
May they crush the egotism and unlimited avarice
that are our mighty foes in this degenerate age!

Above all, may individuals recognize that it is their
greatest personal responsibility to implement the
protection of the world's environment!
Grant your blessings that, having recognized this, from
this day on they can practice whatever they preach,
without voicing mere empty words!

If and when, starting from today, this writer pledges
not to keep tap water running while brushing his
teeth,
Grant your blessings that this may come to pass!

Written by the commoner Khyentse Norbu.

THE GREAT STUPA OF BODHNATH (Tibetan: *Jarung Khashor*) is one of the largest stupas in the world, reputedly constructed during the reign of King Manadeva (fifth century AD) by a poultry keeper named Jadzimo. The stupa, located on the ancient trade route from Tibet to Patan in Nepal, is where Tibetan merchants rested and offered prayers for centuries. When Tibetan refugees first entered Nepal, many settled here. A UNESCO World Heritage Site since 1979, it is now surrounded by some fifty Tibetan Buddhist monasteries. Stupas are architectural representations of enlightened mind, and play a geomantic role in harmonizing environmental energies. *The Legend of the Great Stupa*, a visionary text of the Nyingma tradition telling the story of its construction, contains predictions of the Kaliyuga which are remarkably consistent with the predictive computer models of progressive global warming.

Ato Rinpoche (b. 1933), the eighth Tenzin Tulku, is the nephew of Kyabje Dilgo Khyentse Rinpoche. After leaving Tibet in 1959, the Dalai Lama asked him to direct a monastery for all four Tibetan Buddhist lineages. In 1976 he married and moved to Cambridge, England, where he now lives with his wife and daughter. He worked as a psychiatric nurse at Fulbourn Hospital until taking retirement in 1981. He recently completed the restoration of his original monastery at Nezang near Jyekundo in Kham. He is a master of both Mahamudra and Dzogchen.

Ato Rinpoche

Human Intelligence Without Wisdom Can Destroy Nature

The great snow mountains maintain the health of the rivers of India and China. There is a Tibetan thangka painting that shows Mount Kailash and the four types of rivers it produces: elephant water, peacock water, horse water, lion water. All emerge from the snow mountain and flow down to India, benefiting everybody. In Tibetan tradition, the snow is the mountain's clothes, and it provides great benefit. In my area in Kham there are snow-capped peaks above the meadows, even in summer, and three great rivers flow down into China. They all derive from glaciers that are now receding.

I think everybody including political and religious leaders should discuss global warming among themselves. Everybody has to take responsibility. Everybody needs to work together and then something positive can result. I deeply believe in blessing, prayer, sharing merit, and loving-kindness. In Lord Buddha's teaching, three things are most important: we should develop devotion, meditation, and loving-kindness toward all sentient beings and the whole world. Meditation alone is not enough—these three aspects should work together. This is an effective way to practice in the contemporary world.

Human intelligence has developed all kinds of technologies as the basis of our modern way of life. We consider "development" to be the possession of all kinds of modern technologies. Yet we have also destroyed so much through our polluting industrial and transportation systems.

Question: Scientists tell us we will provoke runaway global warming, if we remain so greedy and ignorant that we provoke climate tipping points... Destroying the Amazon rainforest would be like destroying one lung of our own body. Or we allow the North polar sea ice to melt permanently, so that it cannot reflect sunlight. Were these to happen, we could destroy the very basis of agriculture and civilization, together with half the animal and plant species on the planet.

Ato Rinpoche: I absolutely agree with you about greed. Buddha's teaching speaks of non-attachment. A lot of Western people jump to the conclusion "Oh, I've got a big house, so much furniture, so many books, I must get rid of them." That is not the point. The teaching is pointing out we should not feel we never have enough. Modern technology exaggerates this tendency even more. There is intense competition, even between countries and religions. Nowadays, even religious teachers think, "Why shouldn't I have a Mercedes?"—or even a second one. Greed is a pervasive root cause of our many problems. Non-attachment means we try to accept whatever it is, negative or positive karma. It does not matter if we have quite a lot of money—we can appreciate and enjoy it. The point of non-attachment is we are content, rather than continuously being greedy for more. Human intelligence is corrupted by greed, and that makes human wisdom impossible. Greed stops human intelligence from evolving into wisdom, and human intelligence without wisdom becomes destructive. It can destroy every aspect of nature. It is a matter of great concern that we have already destroyed so much. How are we to handle this developing situation?

Panic is of no use. We might be a good-hearted person with a good mind who wants to join a demonstration. At the outset we are involved in a peace demonstration, but later we lose our temper and there is no more peace. Therefore, we need balance, the middle way. When we are trying between us to bring about positive change, our motivation is very important. When we make a prayer, motivation is essential. Some people say, "Why pray? It's so wishy-washy." Real prayer needs two qualities. It is like hanging a curtain. We need a ring and a hook for it to hang beautifully. We need something we deeply believe in, and the power of blessing. Then our prayer works.

Whatever we do, we need to keep a balance. That is the middle way. Even with simple things. For example, I do not make so many bonfires in the garden now. The whole family tries to recycle whatever we can. This is all about motivation. We change that and there are benefits in unexpected ways. It looks as if just one household does not amount to much change, but we all must try it… especially those who are leaders in society. I teach my pupils not to become guilty. Even if you have done negative or incorrect things, understand that in the future you are not going to repeat them, you are going to do something positive instead. Many people are consumed by guilt, but it just weakens our mind, intelligence, and wisdom. So many people experience a nervous breakdown. We do not have enough patience with ourselves.

Q: Many people are still in denial about global warming. The media have failed to communicate that this is a great evolutionary crisis, and we are creating conditions for our own extinction. If the great mountain peaks have no snow, or the great forests are destroyed, will the human spirit and cultural imagination not be impoverished by such diminishment of nature? It could become evident that the natural world is "ending" in mass extinction. There would then be the danger of collective mental breakdown.
AR: Absolutely true.

Q: One danger of studying this crisis is becoming reactive or angry, destroying the quality of one's communication and peace of mind. What practical advice would you give to meditators?
AR: As spiritual people, our responsibility is prayer. The Four Immeasurable Qualities are included in this prayer:

> May all sentient beings have happiness and its causes,
> May all sentient beings be free of suffering and its causes,
> May all sentient beings never be separated from bliss without suffering,
> May all sentient beings be in equanimity: free of bias, attachment, and anger.

Our Buddhist duty is to all sentient beings, not just to our own species. Without the rest of nature, humans are like fish out of water. We need the water and we need all sentient beings. So how do we handle the situation? How do we gradually get this message out? It is a matter of good motivation. We try to avoid destructive forces that harm nature and sentient beings. I try to point out to my students how beautiful nature is. Many pupils have told me they are lonely. In the *100,000 Songs* Milarepa said, "I am never lonely. I have the deer, the monkeys, the birds. I am so rich." He stayed without food in the mountains and never felt loneliness or nothingness. He said even ghosts were his neighbors.

The message to give people is that nature is the snow mountains and the water is our food and drinking water. So we must put a limit on human greed—we must use human intelligence and human wisdom. This is the function of spiritual teaching. When I was young, my teachers repeated the truths of impermanence, death, and karma daily. They talked about the future of Lord Buddha's teaching, the changing of the kalpa when everything would disappear, and the time when the power of the sun would be seven times greater. We studied these things every day, and now I understand what is happening in those terms.

In Qinghai, Tibet, the beautiful blue lake is diminishing every year and the snow lines are receding. In my sister's well, the level is lower every year. I have been aware of impermanence since my first studies. When Tibet as a country was completely lost, my family went through much suffering. My possessions, my monasteries, my family lands—that whole life is now like a dream. So I really appreciate what impermanence means. Everything is just like an illusion and like a dream. At first I did not truly appreciate this truth. I merely got irritated by my teacher repeating it everyday. Then when all these events really came about, I came to feel the truth of impermanence deeply. We humans never learn things the easy way. We always have to learn things the hard way. That is how it was with me too.

Q: Is there a difference between the classical teachings of impermanence and this present threat, where even the biosphere itself might become impermanent?

AR: It is a different kind of impermanence in that we are the ones who are carrying it out. Human intelligence and human greed are destroying everything—deforestation, too many cars, too many airplanes, too many factories, too many different countries following an incorrect style of industrial development. Furthermore, we waste tremendous resources on militarism.

Q: What about the strong element of tragedy in what is happening?

AR: The question is how do we handle it? It is not helpful to get upset or feel guilty. We do whatever we can with a good heart, with a gentle approach, in a friendly way, with loving-kindness. When it comes to ourselves or others we never forget good work and good motivation. It is not a matter of becoming perfect in one day. We can all make progressive changes, beginning on a small scale. Collectively this can amount to something that changes the world.

Q: Some Dharma practitioners, when they say "all sentient beings," might think "all human sentient beings." If we could extend our meditation to embrace the whole living world, wouldn't that lift our hearts at this time?

AR: Yes. In fact, in the special Vajrayana foundation practices, we visualize the natural environment: a beautiful lake, in the middle of which is the refuge tree. Nature should be an integral part of meditation. For me our response to the current situation comes down to the six paramitas—generosity, self-discipline, ethical conduct, meditation, concentration, wisdom. Whatever special prayer we do, it has to be based in the six paramitas. The positive quality that overcomes greed is generosity. The same applies with self-discipline and ethical conduct. In this way we can offer simplicity, and that leads to general understanding. Nowadays, I teach the simple way. After all that I have studied and experienced in my life, now I go back to first principles, universal principles. Whatever we do we must not forget all sentient beings. Whether we do one prostration, take refuge once, light one stick of incense or a lamp, say one prayer—whatever practice we do, we need to think of all sentient beings, and not only human beings. You might be practicing for something or someone in particular, but in the end you have to have this universal dedication. Buddha gave up responsibility for kingdom, possessions, and family, and practiced until enlightenment. In the end he took on a far greater responsibility, responsibility for all sentient beings.

Ringu Tulku Rinpoche (b. 1952), a reincarnate lama of Rigul monastery in Kham, was a professor of Tibetology in Sikkim for seventeen years. He is well known as a scholar and exponent of the non-sectarian Rime movement in Tibetan Buddhism. His fluent English and congenial teaching style are appreciated worldwide. He founded Bodhicharya, an international organization for the practice and transmission of Buddhist teachings, which also emphasizes intercultural dialogue, education, and social projects.

Ringu Tulku Rinpoche

The Bodhisattva Path at a Time of Crisis

Our pollution of the air and the global environment has created a profound ecological crisis, and it is our human responsibility to correct it. The climate issue is a clear case of karmic action and result. It comes precisely from our incorrect way of doing things—taking action without considering its effect.

When society degenerates, the world becomes worse. Peoples' negative emotions and actions become raw, aggressive, greedy, and deluded. Environmental damage accumulates, militarism and war come to the fore, disease, famine, and diminishing lifespan begin to increase. Excessive greed causes us to disrespectfully take everything from the earth or sea, while ignoring the pollution we cause. This collective negativity harms ourselves, of course, and in the case of global warming, the damage will extend long into the future of ourselves and others. Consumerism relies upon fundamental confusion, amplified by advertising. The resulting over-consumption sows the seeds of self-destruction, as we can now see in the Arctic. The sea ice has melted so much that there is a waterway all round the top of the world. Instead of taking urgent stock of what this may mean for the survival of the world we know, the neighboring countries have started to fight about who gets any oil reserves beneath the ocean.

Great spiritual masters like Padmasambhava saw the world as dreamlike and illusory, yet they went to enormous effort to benefit future generations. Emptiness, interdependence, impermanence, and the dreamlike nature of things do not prevent us from taking altruistic or positive action. It may be like a dream, but it still affects beings. The issue is raised in the *Bodhicaryavatara*—if everything is emptiness, why is there a need for compassion? There is a need because people suffer. They do not understand emptiness. Therefore it is important to work for their benefit, to reduce suffering.

Through interdependence and emptiness, we see phenomena and events clearly. Whatever the situation may be, rather than panicking, we change our way of experiencing. This does not imply we should not try to change the situation. If we have to live in it, we do so in a harmonious and joyful manner. We do whatever we can to make it better—without becoming demoralized or overwhelmed. We choose to live in such a way that we improve things, outside and inside.

There is something we can learn by analogy from the situation of the Tibetans in exile. The Dalai Lama repeatedly told us to remain optimistic. That does not imply ignoring problems or injustices. Neither do we blame ourselves. We see the situation we are in with clarity, understanding, and acceptance, and so we avoid sheer disillusionment. When we find a small potential improvement, we concentrate on that possibility, rather than mourning past or present loss. This not easy, given the numerous negative forces. But it is effective.

So a clear understanding of the scientific information is very important. A few vague ideas about global warming, without any sense of personal involvement, will not suffice. Look at how governments talk a lot, but their inadequate actions fall far short of their rhetoric. There is a Sanskrit verse:

> For the sake of the world, you should sacrifice your country.
> For the sake of the country, you should sacrifice your village.
> For the sake of your village, you should sacrifice your family.
> For the sake of your family, you should sacrifice yourself.

It appears the opposite attitude is prevalent nowadays:

> For the sake of your country, you sacrifice the world.
> For the sake of your village, you sacrifice your country.
> For the sake of your family, you sacrifice your village.
> For the sake of yourself, you sacrifice your family.

When that kind of situation comes about, people think, "If it is somehow beneficial for me, if I get more money for a certain time, I don't care if the planet is going to the dogs." According to this ignorant tendency, our welfare is assured by money or power. Yet we live in this world and if the world is gone, where will we use our "profit"? When we stop believing we are part of anything at all, there is widespread alienation, loneliness, or depression in society. They have the same origin as the climate crisis.

Human Nature, Samsaric Nature, Buddha Nature, End of Nature

The truth of buddha nature applies to all beings. Although the nature of their minds can be profoundly positive and harmonious, they are not like that themselves: they are samsaric beings. Humans are samsaric beings. Our noble potential to be useful to ourselves and others is submerged in our samsaric state of mind. Buddha nature confers the possibility of fundamental transformation. Conversely, ignorance is having the capacity to do something positive for oneself and others, but not knowing how to do so. We tend to call it "the human condition."

Buddhism does not imply humans are good or kind. It identifies a samsaric nature obsessed with the five poisons. As long as we are in this samsaric state of mind, we are bound to react with self-centeredness, affliction, desire, grasping, and basic ignorance. In human nature there is aggression, selfishness, and so forth, but also instinctive empathy and altruism, together with high intelligence. We humans do act aggressively on behalf of self, group, or tribe—especially if we are conditioned to suppress our altruistic instinct. If we understand that aggressive reactions are counterproductive, cooperation and altruism open up, and something superior is accomplished.

Human ignorance sees others as a threat to our survival. In fact, we benefit from seeing everyone as a potential partner. "View" is the most important factor in bringing about this shift. Interdependence is its key element. Everybody wants love and appreciation, and it cannot exist without other people. We can't be appreciated if we try to put down everyone we encounter. Even a "superpower" wants other countries to assist and admire it. But it cannot do so without sharing, learning, and cooperation, or it will come to exemplify collective selfishness and ignorance. This small-minded way of looking at life says to itself, "For me to survive, I need to get the best of others, or if necessary, just get rid of them. I can make best use of the available resources." When it comes to global warming, there is no choice but to take a much bigger perspective. Beliefs like "I will survive but others won't" do not hold up any more. Beliefs like "I'll make an island or stronghold for myself" or "If half the world dies, there will be more for me" do not hold up any more. We have to understand this with complete clarity.

Interdependent External and Internal Realms

The basic Buddhist truth is that everything is interdependent. Our world is clearly something growing, living, and decaying. Eventually it will undergo destruction. Everybody that lives on it is in a "container/contained" relationship—the world and beings. And this living world depends greatly on peoples' mental state, how we experience our reality.

Innumerable problems come from our failure to understand that phenomena are impermanent. We plan as if we are going to live five hundred years. We fight about little things. People seek power and hold on to it, imagining they will have it for generations. Even if they get it, nothing lasts. Conflict, aggression, failure to accommodate others' feelings or live in harmony all follow from the misunderstanding of impermanence. When we do understand it, we also understand interdependence. These two have a subtle balance, and if we cannot maintain that subtle balance, phenomena rapidly become disordered. This is what we see with the whole climate and environmental crisis. Things are not only constantly changing, but delicately balanced—in a subtle way that could break at any time. Little things can have a big effect. Furthermore, impermanence does not only mean things will fall apart. Impermanence means things are not as solid as we imagine. They can get better or worse. They are changing, interdependent, and easily affected.

Since ignorance is our basic problem, truth leads us to selflessness and the deep understanding of the nature of things. This is a process of self-understanding. Emptiness and selflessness cannot be left at the level of words. The selflessness of things does not mean they are "not there." It means that everything is interdependent. Selflessness, emptiness, impermanence, and interdependence—these are terms that indicate how to discover inner peace, joy, and contentment. On the collective level, they point to the Buddhist social model of gross national happiness, rather than greedy consumerism and its side effects of competition, dissatisfaction, anxiety, and pollution. How rich we are does not depend on how much we have, but on our level of contentment. If we feel we never have enough, we are poor even if we have the whole world. We become like "treasure-god hungry ghosts" who sit and guard a big treasure behind a locked door. Unable to enjoy it out of anxiety that someone will take it, we are stuck in a hungry ghost realm. Genuine wealth is to feel we have enough.

Now, during a "degenerate time" the kleshas, the poisonous mental states, become collectively strong and intense. Buddhist practitioners train for this by preparing for the worst. We prepare for our death as a main practice. We train not to lose the calm clarity of the mind even in the worst situations. Of course, there are those who assume we should just sit in bliss and peace, that ourselves, not only on the external dimension. The worse the situation in the world gets, the stronger our spiritual practice could become. External chaos becomes an eye-opener. It demonstrates the futility of chasing worldly things that yield nothing worthwhile anyway. As we work on a different level, changing how we experience this life, there is an influence on other people and the possibility of transforming the outer world arises. If I really change my way of experiencing things, however bad the outer world becomes, it does not affect me in the old way. I experience its interdependent appearance-emptiness. Whatever happens, I let it be and relax. Nothing appears "wrong" anymore, and ultimately nothing in this world, nor in the bardo, the intermediate state between incarnations, can overwhelm me. When I

y body but cannot
eration.
nal circumstances do not
y can be greatly affected. We
nd liberation, but we should also
ollective, negative phenomena do not
lves are beyond being affected. Being in
nally does not mean that others experience
body has their own mind stream. The collective
the situation also exists. If there is environmental or
ollapse, everybody will assuredly be affected—some
some less, but there will be an unprecedented negative
pact. Clearly it is a vitally important bodhisattva activity to
prevent a universal disaster like the collapse of our living world.

FLOATING MARKET IN THE MEKONG DELTA, VIETNAM: The Mekong is the seventh longest river in Asia and exemplifies the vulnerability of Asia's cultures and peoples to global warming. Running from the Tibetan plateau through China, Burma, Thailand, Laos, and Cambodia to the sea in Vietnam, its basin is one of the world's richest areas of biodiversity, and it contains over 1,200 species of fish. Tibetan glacial and snow melt provides steady year-round flow at the river's sources, but this is vulnerable to rapid climate warming on the plateau. The delta rice paddies of Vietnam, vital for food security and export markets, are only one to five meters above sea level, and are now endangered by sea level rises predicted to result this century from current carbon dioxide emissions trends.

Chokyi Nyima Rinpoche (b. 1951) is the head of Ka-Nying Shedrub Ling Monastery at the Great Stupa of Bodnath, Kathmandu. He is the eldest son of the eminent Dzogchen master, Tulku Urgyen Rinpoche. An internationally celebrated meditation teacher in his own right, he founded the Rangjung Yeshe Institute for Buddhist Studies in 1981. His published works include *Union of Mahamudra and Dzogchen*, *Bardo Guidebook*, and *Indisputable Truth*.

Chokyi Nyima Rinpoche

Very Dangerous Territory

We human beings are greedy. On one hand we feel smart because we are very civilized, we have so much technology, and we have become rich and comfortable. A few thousand years ago, people lived hard lives. There were no airplanes, cars, or even bicycles, and travel was difficult. Worse, there were no modern medicines, not even proper eyeglasses! Now we are attached to our comfort and our high standard of living. On the other hand, we are just burning through everything, expanding our population, and using up the Earth's natural resources. It's unbelievable. We are actually just living greedily in short-term luxury—but the way we live now will affect everything for a long time. The damage will not be minor at all; it might assume such proportions that life will end up much worse than it was a thousand years ago.

Humans are getting into very dangerous territory. The children of the future will face great difficulties and danger—shortage of water, food, medicine; shortage of everything. This is a very real danger, not only for humans but also for all the animals. Our world could turn into a desert and we really need to be aware of this.

Right now, we are in the middle of an emergency. Every day we are falling further behind and remedial actions are getting postponed. We shouldn't wait years, or even months, because each day we are causing further damage to ourselves. We can no longer think simply about Tibet, Asia, and America since this crisis is assuming global proportions and affecting the whole planet.

Now, primarily, this is in the hands of politicians. But we also need to have religious leaders, scientists, and other relevant experts discuss this issue and come to agreement as to how to best solve this problem. We need the leaders of all religions to be involved. It's very important that religious leaders, politicians, business people, and industrialists have a genuine dialogue now. I think everybody will care, if not for other reasons, then at least for his or her descendants. This is about our own children, grandchildren, and their sons and daughters.

We need to have a serious discussion and get the general public involved. Everyone needs to know the direction that we are heading. I think the main point is to focus on education and access to information. We need to know what kind of damage we are creating and what can be done to reverse that. What is the solution? Expert scientists now say there is less than ten years left to make an effective world treaty and put it into practice. The well-known Buddhist text, the *Abhidharmakosha*, implies the future world will become very hot—seven times the heat of the present sun.

Buddhists believe in the effect of aspirations and meditation. It is very important that all religions make their prayers and perform rituals to turn around this global warming situation. If religious practitioners begin to focus on global warming in such ways, then others may also notice it and begin to enquire into the nature and purpose of these rituals and prayers. People will pause and begin to contemplate the current situation. People will become more aware, and this will have a positive influence.

Religion may be the only motivational force as powerful as the big business interests that are blocking constructive change. Buddhist leaders could talk to scientists and reach a common position. It is absolutely not acceptable to destroy the future for our descendants and for all these other species. We absolutely condemn it. We absolutely, structurally, oppose it. Naturally, that would be the Buddhist response. I am not joking.

My conclusion is that we need to create awareness about this among all human beings. It is not possible to explain this to the animals but, if it were, we should even do that. We need all human beings to recognize the implications and outcomes of our current lifestyle. We have become so greedy that we choose today's luxury over any consideration of future consequences: "What happens tomorrow, happens… I just take care of my life. Who cares what kind of life my children will face?" We are actively destructive, but we don't care. Above all else, we want our comfort now.

So it is urgent to clearly demonstrate what effects will appear in the future, as well as the solutions available now. We need to encourage religious leaders, politicians, business people, and the general public to look at these trends and outcomes. Everyone needs to focus with a common energy on the available solutions. What is the best solution? How important is it to apply it? Why not apply it now? If we don't apply it, what is the danger? We need to move to the solutions. I can't say more than that.

Tsoknyi Rinpoche (b. 1966) is recognized as the reincarnation of Drubwang Tsoknyi, a renowned master of the Drukpa Kagyu and Nyingma traditions from Nangchen in Kham. His teachers were Khamtrul Rinpoche, Kyabje Dilgo Khyentse Rinpoche, Adi Rinpoche, and his father, Tulku Urgyen Rinpoche. He is abbot of two nunneries and one monastery in Nepal, as well as one of the largest nunneries in Tibet. He also heads some fifty practice centers and hermitages with over two thousand nuns and nine hundred monks in Nangchen, Tibet. A renowned teacher of Dzogchen meditation, he is the author of *Carefree Dignity* and *Fearless Simplicity*.

Tsoknyi Rinpoche

A New Meaning of *Chu* ("Beings") and *No* ("Environment") Has Emerged

When people here in Nepal hear about global warming and energy supply, they just want to know who will provide the solutions. They live hand-to-mouth, so where are they going to get any alternative? Their attitude is, "You will have to supply us with it. You make the changes and then we will use it. My grandfather didn't use lots of energy. My father didn't use it. Now I'm using a little bit but you are making such a noise about it—so bring us alternatives and we will use them."

When I talk to people here about global warming, so far the response has been aggressive. Some Indian people get angry—to them, it's like the nuclear issue: "You people already have nuclear weapons, but we can't make a deterrent. You have a hundred times more." Poor people resent being advised not to cut wood or use kerosene. "So how do we cook our food? The fire will not come out of thin air. You say we cannot cut wood because it is bad for the environment. Now you say avoid kerosene too, that's also bad." Over here we're talking about today's food level—survival rather than choice or comfort.

Government people here think like this: "We didn't use as much fossil fuel, and now we hear 'Let's share the suffering.'" If you want to educate people in this part of the world, you need to give chocolate first. If you invite an expert from America to speak about the environmental crisis, nobody will come. People think, "Will we get food or tea? Will we get a special pen or pencil? OK, then we'll go." Or they think "We are facing such day-to-day living problems, why should we bother with an environmentalist's lecture?" You always have to give something in order to transform something here. In the West you pay for such things. Here it is completely different.

In the case of global warming, everyone is going to suffer, and we know there could be great scarcity of water and agricultural problems. Maybe a positive plan could emerge from the developed nations that would benefit people here in their day-to-day lives—something affordable and accessible, like fuel-efficient stoves.

When it comes to any renewable energy project here in Nepal, someone external needs to be responsible for completing the whole job. If it's done in pieces, if you give money here and there for parts of the job, nothing will happen—someone comes to put up the support poles and a few months later they fall down and you can't get anybody to take responsibility. As a result, people here only have confidence in a project if a foreign organization is involved and does everything. If money is poured into the local system, it just disappears.

Deep down every Buddhist practitioner would like to live peacefully. Our appreciation for the natural world is very strong. You cannot have a peaceful mind if the environment is unstable. It is all related. There could be many Buddhists who would have a deep interest in setting up renewable energy programs over here. And that would certainly help our basic economy. We are already paying a lot for electricity, and it's increasing by 15 percent every year. I feel Buddhists would support reducing this financial drain while helping the environment—through a well-designed carbon-offset scheme, for example.

Talking to the government here is a waste of time, but setting workable goals for individual monasteries could make renewable energy solutions happen. If you could make it happen at one monastery, using a single company for everything, people would start to wake up. That could change attitudes in government. It would also be worthwhile to discuss this with young, influential lamas. After all, global warming isn't an ordinary political matter. Scientists have found it happened previously in the early history of this planet. It is a universal problem and the possible outcomes are extremely negative for humans and other species. We should be concentrating on the solutions.

Tibetan Buddhism has many limitations because everything is newly re-established. We had to just run out of our homeland. Everybody was concerned with making a new base. We still cannot settle down in one place, and call it our own. We stay in Nepal, but here too, there are significant political problems. Everyone's mind is fully occupied trying to establish some basic stability and development. When we hear about a big global issue like this, it doesn't really click yet. So if many senior Buddhist teachers seem not to have thought about this issue, their feeling might be "It's samsara, and samsara always has this kind of thing happening." They are not clinging and that is beautiful, but there is an element of shortsightedness if one doesn't think ahead. Some teachers are busy supporting monasteries. Others understand it perfectly. I myself can understand it, but my "front page" is the monks' and nuns' survival. First I must take care of their food. For example, I have just spent 200,000 rupees for a sick monk to get hospital treatment here in Nepal. I felt happy for that person, but later I reflected that with that much (around $3,000) I could have helped more people in Tibet. I have to arrive at a place where the basic survival needs of the monks and nuns are taken care of. The monastic sangha is my direct responsibility, and many lamas face this same problem. Whatever money we receive is for survival—food, clothing, blankets. Nothing is left over.

If you ask whether Buddhists should work for solutions to global warming more on the spiritual or physical plane, both are equally important. Spiritually, from the Buddhist point of view, whenever we pray, we pray to balance *no* and *chu*. *No* means the container, universe, or environment; *chu* means what's inside, the beings. Our prayers include aspirations for averting natural disasters due to an imbalance of the elements of earth, water, fire, air, and space. They also emphasize reducing the poisons of ignorance, hatred, desire, pride, and greed, while increasing the virtues of loving-kindness, compassion, and wisdom.

A root concern for the environment is already in our prayers. It is part of Buddhism, in the teaching of interdependent origination. Lamas have traditionally balanced the elements in the environment by building stupas, planting treasure vases, blessing the land with *rab-ne* ceremonies, raising prayer flags, making *tsa tsas*, etc. We have worked hard in this way, from the time of King Ashoka and Lord Buddha. Now *no* has taken on a new meaning and we are facing a modern, sophisticated challenge. Before, imbalances in the environment were due to natural causes. Now major environmental problems are being created by humankind itself. It takes some time to understand the causative factors fully. Averting global warming will require new education and new understanding. We will need new prayers. I am sure, after we fully understand the issues, Buddhists will come to the forefront and work not only on the spiritual level, but on new, physical solutions. I am very much concerned about all of this. I will pray, and I will try to influence whomever I can reach—especially monks and nuns.

Dzigar Kongtrul Rinpoche (b. 1964), is recognized as a recent emanation of Jamgon Kongtrul Lodro Thaye (d. 1899), with particular affiliation to Dzigar in Jomda in eastern Tibet. Born in exile in northern India, he received spiritual guidance initially from his father Neten Choling Rinpoche of Nangchen and from his root teacher, Kyabje Dilgo Khyentse Rinpoche, through whom he holds the Nyingma Longchen Nyingthig and Rime Khyen-Kong-Chok-sum lineages. He moved with his family from India to the U.S. in 1989, and founded Mangala Shri Bhuti, an organization integrating several retreat centers. He spends time in regular retreat and at the monastic seat of Neten Choling (Himachal Pradesh, India), as well as teaching throughout the world. He is an abstract painter and the author of *It's Up to You* and *Light Comes Through*.

Dzigar Kongtrul Rinpoche

Minimum Needs and Maximum Contentment

For anyone wishing to live responsibly in this world, we have to address global warming. If it is not addressed now, finding a solution will only become more difficult with time. If we act too late, it will be disastrous. We have no other option but to think seriously about sustainable living on Earth.

Human beings, other species, plants, and trees exist together on this planet. The living world, atmosphere, and climate share an interdependent relationship, as is becoming increasingly apparent. Our actions have a much broader effect than on our lives alone. We must increase our sympathy for the well-being of other species and ensure their continuation on this Earth. The same holds true for plants, trees, even air and water—they are all living environmental factors. When one deteriorates, others also weaken. Awareness of interdependence shows us that when plants become extinct, that means the loss of invaluable human medicines. When forests shrink, that means the loss of climate regulation. Understanding and guarding this interconnectedness is one definition of "virtue."

Energy is a primary issue in the debate about sustainable living, for current and future generations. The energy sources to which we are accustomed have become problematic for the environment. We have an especially unhealthy dependence on fossil fuels. When oil runs short, what are we going to do? Particularly in times of crisis, reality can appear fixed and unchangeable; we might believe that we have limited alternatives to fossil fuel or those we do have are too difficult to develop. But from the Buddhist point of view the ultimate truth of the world in which we live, the Dharmadhatu, is free of any fixed reality. Therefore it could be a source of multiple possibilities. Given that understanding, as Buddhist practitioners we can make supplications and seek to accumulate collective merit for sustainable, alternative energy solutions to manifest. If there is aspiration and merit, I believe energy may be discovered from as of yet unrecognized sources.

These times require that the powerful special interest groups who hold to traditional energy models undergo a change of heart. To do so they will have to confront their own greed. Greed is not easily changed. Throughout the developed world the karma of greed is ripening in a pervasive manner. At the same time, virtue is appearing in front of all of us. Collectively we are becoming aware that things have to change. We can see that choice clearly now. Instead of affecting indifference, Buddhists should reflect deeply and act appropriately. We can make strong aspirations for positive change to limit global warming before it is too late. Thinking negatively, blaming others, or feeling hopeless will not help. A hopeful attitude, and taking action, will bring rmation. As a citizen and also as a Buddhist, we gather our s, create dialogue, and promote positive change. This raises consciousness of what is happening. Consciousness is very powerful. It is actually the most powerful thing. We have to do what we can to enlighten the current situation.

We must publicly acknowledge that our actions are causing tremendous destruction in the world; this would be a positive step forward. The classical Greek model of the republic recognized that the advantaged class could make selfish errors that did not serve anyone's interest—their own included. A republic cannot exist solely for the strong and advantaged. It is the responsibility of the general population to question the elite, and that is what is called for right now. We are in a critical situation, and my feeling is there will be fundamental change, because everyone has a stake in what is in front of us. Everything in the future depends on the virtue of people alive here and now, on whether human beings can relate to each other and the environment more harmoniously.

The capitalist economy and attendant consumerism that dominate First World cultures breed a compulsive feeling that we don't have enough, even when our closets are full of everything and anything. That is the effect of half a century of television advertising. What Buddhist teachers and practitioners can do now is demonstrate how we can break away from this culture of addiction. This will take some strength, because we too can be quite immersed in consumerism. We will need inner discipline to break from it and choose a lifestyle of minimum needs and maximum contentment. Now this is crucial to understand for many reasons, including the health and integrity of our own mind stream. If we cannot accomplish it, progress on the spiritual path will be difficult.

This one arrow—minimum needs, maximum contentment—hits a number of important targets. It helps the environment; it is a constructive model for others; and it supports our growth on the spiritual path. We should contemplate it in depth, and take steps toward sustainable energy and lifestyle now.

On the whole we find that the environmental movement shares many priorities with our way of life, and lends more credibility to the Buddhist path. The environmental emergency brings up many key issues of consciousness for discussion. Society in general is now beginning to consider the Buddhist principle of a simple life. For example, the life of someone on retreat demonstrates the human capacity for contentment with very little. As a model, this discipline of minimum needs, maximum content speaks directly to the type of individual conservationism that is a theme of environmentally conscious culture. Each person whose choices are not motivated by the capitalistic impulse but rather by a spirit of self-sufficiency can have tremendous impact.

This is not to say Buddhists may not themselves be quite enmeshed in modern consumer culture and technology. So many optional things have become convenient "necessities" that are difficult to relinquish: our car, washing machine, computer, and so forth. In my early life, I didn't have the money to purchase a refrigerator, let alone a washing machine, central heating, or a car. I consider myself fortunate to have had to use public transportation. When we used electric heaters, my mother used to tell us, "That costs a lot of money and uses a lot of energy. Don't use it." If we went to sleep and forgot to turn off the light, the next day we had a big lecture about waste.

We Tibetan refugees were poor and couldn't afford those things. In point of fact we were living in harmony with nature and in balance with the environment—though we would not have recognized it as such. Now one sees all the problems consumerism has caused for the environment while people work round the clock for all these "conveniences." I much appreciate having grown up without them. We should educate people in the Third World to understand that it's not poverty to live outside the confines of consumer society. In fact it is richer for the human spirit and the environment. This is a truth that must be expressed by teachers with connections in developing nations.

In the twenty-first century, we should be able to use science and technology to create a waste-free, sustainable world for humans, animals, and plants. This is the highest, and as of yet undiscovered, aspiration for science and technology. Until now, capitalism has pirated science and technology. The great majority of humans have not encountered the higher awareness that science can provide. They live with a limited perspective on the environment, their own species or their own well-being. Businessmen and corporations make profitable use of science and technology, while academic scientists communicate with fellow specialists but cannot reach the general public. There continues to be a big gap between what they know about the climate crisis and the public understanding of it.

I'm cautiously optimistic that the rise in oil prices has led to a period of serious questioning about our lifestyle. The next five years will be crucial. If we cannot get serious about the change from fossil fuels to renewable energy, it's going to be truly difficult to handle this problem in the future. People must rise to the occasion of changing the government. The larger and more prosperous countries suffer from poor leadership on these issues and complicated, dependent relationships with corporations. Meanwhile the tax system does not distribute excess corporate profits to public healthcare or other important areas of society. If you have a hundred thousand citizens on one side of an issue and ten on the other, whom should government support? Yet the hundred thousand are unable to rise up and take power. Ten skillful manipulators with a lot of media influence can condition the majority. We need to do something different, something inventive. We need to speak the truth.

As I see it, the two main weapons of the environmental movement are education and ethical taxation of waste and carbon emissions. Education articulates the issues and ethical taxation puts them into practice. Between the two, we can make the change from fossil fuels to renewable energy a reality. Our MBA programs must teach philosophy as well as economics, statistics, or how so-and-so won this year's award for marketing. They need to address critical global issues, and start a trend that seeks to make products renewable and sustainable. Furthermore, we need a general higher education system that ensures students are aware of these issues. Education is a counterforce to corporate greed. Finally, we need a mass movement to pressure our governments to change the tax system so that it factors in the destructive power of carbon emissions and the associated costs of reducing greenhouse gases. The fossil fuel industry does not want to yield power to new, future energy sources. We could give them a choice—either invest in renewable energy for profit, or the government will create a "windfall tax" on their record profits, and then invest in renewable energy on behalf of the country.

All religions are based upon principles that constitute an ethical way of life. Our current lifestyle does not uphold the human
[illegible] does not support an ethical way of living. In response,
[illegible] around the world must advance a strong spiritual
[illegible] to climate change based upon common principles of an

ethical way of life: We are not against wealth or business in themselves. We simply point out that that it's good for business to lose a bit of that excess weight. It's good for business to make a positive contribution to the world. And it's good for individuals to examine their own consciousness in terms of what sustains them while living on this planet.

While corporations continue to pretend climate change is nature's doing, those who lack such self-interest admit that global warming would not be a problem without ongoing fossil fuel use as a cause. The karma of global warming is not nature turning against us—we have turned against ourselves. We are doing something hostile to nature. It is not that "God has turned against us"—Hurricane Katrina was a manifestation of global warming. If we wish to avoid such disasters, we have to take corrective measures now. Our climate itself is now in our own hands.

Part IV
Western Buddhist Perspectives

Bhikkhu Bodhi is an American Buddhist monk. He has edited and authored key publications in the Theravada Buddhist tradition. He holds a doctorate in philosophy, and was first ordained as a novice monk in the Vietnamese Mahayana order. He later traveled to Sri Lanka where he received full Theravada ordination under Ven. Ananda Maitreya in 1973. In 1984 he succeeded Ven. Nyanaponika Thera as English-language editor of the Buddhist Publication Society and in 1988 he became its second president. In 2000, he gave the keynote address at the United Nations' first official Vesak celebration. He currently teaches at Bodhi Monastery (New Jersey) and Chuang Yen Monastery (New York) and is chairman of the Yin Shun Foundation. Among his publications are *The Middle Length Discourses of the Buddha, Numerical Discourses of the Buddha, The Connected Discourses of the Buddha*, and *In the Buddha's Words.*

Venerable Bhikkhu Bodhi

The Voice of the Golden Goose

One of the Jataka stories, the *Palasa Jataka*, recounts a past life of the Buddha when he was a golden goose living in the Himalayas. The goose would regularly stop to rest in a large Judas tree, where he befriended the resident deity of the tree. On one occasion a certain bird, which had eaten the fruit of a baṇyan tree, perched in the Judas tree and voided its excrement into a fork in its trunk. A seed of the banyan fruit thereby germinated and began to grow into a small banyan tree.

When the golden goose next visited the deity in the Judas tree, the banyan had grown four inches and had bright red shoots and green leaves. The royal goose told his friend: "Every tree on which a banyan shoot springs up is destroyed by its growth. Don't allow this banyan to grow in the fork in your trunk or else it will destroy your home. Remove it immediately before it tears your dwelling limb from limb."

The Judas deity, however, demurred: "The shoot is small and harmless. It will provide shade and delightful tendrils." The goose insisted: "The shoot is dangerous; it will lead to your harm; as it grows bigger, it will push you off your tree and even destroy your home." When the Judas deity persisted in its decision, the goose understood it was futile to press its argument. It thus flew away and never returned.

As time went by, all happened as the golden goose had foretold. The banyan shoot sent down roots which wrapped around the trunk of its host and consumed its share of soil water and nutriment. The banyan grew bigger and stronger, until it split the Judas tree, which toppled to its death, bringing the deity's home down with it.

This ancient tale, originally intended as an allegory for the destructive power of evil, can be read as a parable for our present-day crisis of global warming. The banyan seed represents the use of carbon-based fuels, whose emissions of carbon dioxide are invisible, odorless, and, in small quantities, harmless. Just as the banyan shoot grew slowly and imperceptibly, so these fuels, diffused into the atmosphere, produce changes that are initially undetectable. But just as the mature banyan tree destroyed its host, unrestricted use of fossil fuels menaces the civilization that depends on them.

When coal was first used to transform water into steam and thereby drive a steam engine, no one could have foreseen that this marvelous invention marked the beginning of a trajectory that would one day even threaten the prospects for human life on Earth. Yet this is precisely the predicament that we currently face. Like the deity in the Judas tree, we have had sufficient warnings: many golden geese have cautioned us about the dangers of excessive carbon emissions. Already in June 1988 a scientific conference in Toronto on climate change concluded:

> Humanity is conducting an unintended, uncontrolled, globally pervasive experiment whose ultimate consequences could be second only to a global nuclear war… It is imperative to act now.[1]

In the same month, James Hansen, director of NASA's Goddard Space Center, told Congress that it was virtually beyond doubt that burning fossil fuels was warming the Earth. He warned that if current trends continued the result would be more severe droughts and heat waves, but also heavier rains and more frequent floods.

Over the following two decades, successive American administrations failed to heed these warnings. The past eight years of the Bush Administration have been particularly disastrous for efforts to curb emissions. Despite uncanny climatic disasters, from drought in Australia to more intense hurricanes in the southern U.S. to relentless floods in Europe, despite the overwhelming testimony of unbiased scientists, the White House has offered only repeated prescriptions for failure. Administration officials have even muzzled government scientists and censored their warnings that global warming is indeed a consequence of the use of fossil fuels.[2]

When he again appeared before Congress, on June 23, 2008—twenty years to the day after his initial testimony—Hansen stressed that time is running out:

> We have used up all slack in the schedule for actions needed to defuse the global warming time-bomb. The next president and Congress must define a course next year in which the United States exerts leadership commensurate with our responsibility for the present dangerous situation. Otherwise it will become impractical to constrain atmospheric carbon dioxide, the greenhouse gas produced in burning fossil fuels, to a level that prevents the climate system from passing tipping points that lead to disastrous climate changes that spiral dynamically out of humanity's control. Changes needed to

> preserve creation, the planet on which civilization developed, are clear... I argue that a path yielding energy independence and a healthier environment is, barely, still possible.[3]

Hansen defined the precise targets we must meet to prevent warming from reaching catastrophic tipping points. The safe level of carbon dioxide in the atmosphere, he stressed, is no more that 350 ppm. What makes our current situation particularly ominous is that it is already 385 ppm and rising about 2 ppm per year.

The prospects for the future, however, are not encouraging. Demands worldwide for electric power, cars, and meat (a major source of carbon pollution) are escalating. If higher carbon concentrations should push the average temperature of the planet 2°C above pre-industrial levels, this could trigger the irreversible melting of the Greenland ice sheet, a process that has already started. The melted ice would cause a dramatic rise in sea levels of at least four to six feet (about two meters) by the end of the century, flooding the coastal belts of all continents and inundating all major coastal cities. Altered climate patterns would usher in severe droughts, resulting in crop failures, famine, and possible mass starvation. Desperate battles could erupt over dwindling supplies of food and other resources, and vast populations would migrate in search of food. If the mountain glaciers of the Himalayas, Andes, and Rockies disappear through climate warming around mid-century, the billions of people who depend on them for fresh water would be left dry and desolate. Violent conflict over vanishing sources of water could replace the "oil wars" of the present.

Reckless human activity not only contributes directly to global warming, but also triggers feedback processes that accelerate the buildup of greenhouse gases in the atmosphere. Three such feedback loops have already been instigated. There is progressive and self-generating loss of the "albedo effect," the benign reflection of sunlight back into space by white snow and ice. There is deforestation, which releases large quantities of carbon dioxide, removes a major "carbon sink," and by further raising temperatures, provokes more forest fires. A third feedback process involves prodigious frozen deposits of the highly potent greenhouse gas methane. These have begun to be released from Siberian (and Tibetan) permafrost, and even from below the Arctic Ocean sea-bed. They might create a giant positive feedback loop leading to more permafrost melting and methane release.

Such is the predicament that we face today: a planet in peril, moving ever closer to what energy expert Joseph Romm has called "hell and high water." What can we, as Buddhists, do to ameliorate the crisis of global warming and thereby avert the calamities that may follow if urgent, effective, and earnest action is not taken? Is our situation beyond redemption, or is there still room for hope?

As a spiritual teaching, Buddhism rests on two complementary pillars, wisdom and compassion, both of which can help us diagnose and address the dangers of climate breakdown. Through wisdom, we investigate a danger: see it as a whole, identify its underlying causation, and determine what can be done to remedy it at the causal level. Through compassion, our hearts feel the danger vividly and personally, and thereby expand to embrace all those exposed to harm: all who, like ourselves, are subject to suffering, who seek peace, well-being, and happiness.

Reflection on the broad consequences of runaway global warming enables us to see that this is not merely a problem of rules and regulations that can be solved by a simple technological fix. It is at base a deeply moral problem that challenges our humanity and ethical integrity. The fact that billions of human beings on this planet, as well as countless forms of non-human life, have to bear the brunt of the misery caused by the irresponsible behavior of a small number of nations—those that contribute most to global climate change—presents us with an ethical crisis that sears our conscience. This is particularly the case because the populations most likely to be hit hardest by the effects of global warming are those already living in poverty: the people of sub-Saharan Africa, where droughts will get worse; the inhabitants of Central and South Asia, where crop yield could drop by 30 percent; the populations whose island-nations would be swallowed by the sea; and the residents of the Asian mega-deltas, where billions will be in danger of floods. Additionally, if temperatures rise between 1.5–2.5°C, a quarter of all plant and animal species are at risk of extinction.

If we let our minds embrace all our fellow beings with loving-kindness, "as a mother does her only child" (Metta Sutta), we will feel a compelling sense of urgency swell up from our depths, rooted in a clear recognition of the perils that hang over innumerable beings, human and non-human, whether in this country or in other lands. And if we let our hearts be stirred by compassion, we will see that we have no choice but to act, and to act in ways that will truly make a difference. But effective action must be rooted in insight, in wisdom. Here the heuristic approach of Buddhism becomes pertinent. To resolve any problem effectively, it is necessary to see it whole and in its wider context.

As we grope for a solution to global warming, it is worth exploring the question: "What prevents the world community from adopting measures to curb carbon emissions rapidly and on the required scale?" If a fire breaks out in my house, I would quickly take any action necessary to extinguish it, even calling the fire department if it gets out of hand. Yet oddly, when our planetary house is aflame, we spend more time squabbling over who should extinguish the fire than we do actually pursuing ways to put it out. Those most responsible for setting the fire in the first place scheme and bargain to avoid making a full commitment to firefighting. The international protocols and agreements proposed to control carbon emissions, including the Kyoto Protocol, have been weak, limited, and flawed. In the U.S., proposals to Congress to impose mandatory limitations on carbon emissions and other controls on energy consumption have repeatedly failed to receive widespread support. The White House has consistently opposed regulating carbon emissions, contending it would cost too much and hurt economic growth.

Why is there this procrastination and denial, this reluctance to take the sweeping steps needed to save human civilization from almost certain calamity? Why has the compelling consensus of the international scientific community been punctured by the doubts of politicians who rely on the opinions of people outside the scientific community? Why do these deniers, at obvious risk to themselves and their children, sow seeds of skepticism among the general population?

A partial answer arises from an understanding of the strong grip that greed has upon the human heart, and when the people at the helms of the oil, coal, and gas companies—those most responsible for emissions—act in the grip of greed, we can see how immense is the resistance to effective controls on emissions. Economic power does not operate in a domain of its own but is intimately interwoven with political power, and in the U.S. the ties between the two are extremely hard to break. Through their lobbyists, the carbon energy companies have a substantial impact on the formulation of public policy, and the reluctance of politicians to press for stricter controls on carbon emissions is almost directly proportional to the contributions they receive from these industries. This is the hurdle we must clear if we hope to institute a sane environmental policy.

In the public sphere, ecology is locked in such a tense wrestling match with the economy that the relationship between the two seems to be one of inverse advantage. According to the dominant economic model, for an economy to thrive, it must become more productive, turning out a greater number of goods and services. Enhanced productivity involves increased use of energy, and since the energy comes from electricity (provided largely by coal plants), and goods must be transported (with oil-derived fuels), a successful economy almost inevitably results in higher carbon emissions. In the framework of our present economic model, placing restrictions on carbon emissions means limiting productivity, and a decline in activity entails economic losses that will spread throughout the entire economic order. Limits on productivity would usher in a recession, layoffs, lower wages, and reduced employee benefits. Within this framework, the one escape from this fate is increased production, which brings us back to square one.

While unmitigated greed certainly contributes to the resistance corporate leaders show to proposals to curb carbon emissions, greed alone is not a sufficient explanation. Buddhist psychology teaches that greed often coexists with a strong impulse to dominate and control, and this seems pertinent here. For the executives of the oil and coal industries, control over the mainsprings of the economy, its reservoirs of energy, confers the exhilaration of power, of knowing they can dominate national and international affairs. This intoxication with power is hard to relinquish, even though unrestrained use of these fuels endangers the planet, including the executives themselves and their descendents. Another motivating factor may be fear of stagnation, decline, or even collapse, if productivity is curtailed due to controls on carbon emissions. Such fear increases when, as recently, major financial institutions and automobile manufacturers totter at the edge of survival.

Another explanatory factor, most prominent among the general population, is delusion. Delusion screens from our minds the imminence of danger, giving us a foolhardy sense of invulnerability in the midst of insecurity. In our private lives, we unconsciously assume that we are immortal beings, exempt from old age, sickness, and death. Delusion also plants in us the assumption that our environment will always remain safe and secure. If hurricanes, droughts, and floods should strike, we persuade ourselves that they are only temporary displays of the moodiness of the weather and console ourselves that things will "return to normal." We resist acknowledging that our own behavior can be responsible for a potentially catastrophic and irreversible transformation of our climate. Such delusion, already entrenched in the human mind, is strongly reinforced when the energy corporations use their wealth and influence to spread disinformation and cast doubt on the truth and urgency of global warming.[4]

Thus, in the U.S., denial of global warming and indifference toward its consequences stems from the potent mixture of corporate greed, the arrogance of power, fear of collapse, and stultifying delusion. In developing nations, in contrast, reluctance to counter global warming often springs from the sheer struggle for survival. These countries see enhanced economic activity as essential to their escape from poverty. Thus cuts in carbon emissions, in so far as they hamper economic productivity, appear to them as an unwelcome impediment to their aspirations for greater prosperity. From their perspective, environmental concerns are of limited importance; many in these countries even regard environmental issues as a peculiarly Western obsession. For the population of the developing world, the chief priority is to emerge from endemic poverty, and the road to this goal is economic development. A robust economy is one that constantly produces more goods and services, and thus necessarily consumes more energy. To compel such economies to impose caps on carbon emissions is, from their own point of view, to consign them to continued poverty, a bitter fate with which they are already too familiar.

Nevertheless, even the most "productive" economy cannot flourish on a planet beset by a torrent of environmental woes stemming from runaway climate change. Now that major ecological disasters have already started to manifest, our only chance to avert a catastrophic fate is to promptly adopt sweeping and comprehensive measures to reduce global warming. These must apply equally to both the developed and developing worlds.

Major changes are needed in both our personal conduct and our economies. While in our private lives we should certainly increase our environmental awareness and adopt as many small behaviors as possible that reduce our personal carbon footprint—for example, retrofitting windows, using compact fluorescent light bulbs, turning off unused electric appliances, driving a fuel-efficient car, eating mainly vegetarian, keeping the thermostat low—it would be overly optimistic, even naïve, to believe that merely reducing our personal share of carbon waste and encouraging others to do so will bring emissions down to the levels needed to avert future calamity. Global warming is a problem of truly global dimensions, intricately related to the gargantuan quantities of energy our economies devour to sustain forward movement and support the standards of living we have learned to expect. Hence there is no alternative to a wide-ranging plan that will dramatically transform the entire economy—our own and that of the world.

From a Buddhist perspective, a sane economy would be governed by the principle of sufficiency, which holds that the key to happiness is contentment rather than an abundance of goods. Our unquenchable compulsion to consume and enjoy is an expression of craving, the very thing the Buddha pinpointed as the root cause of suffering. In the dominant market-based model of the good life, the economy is governed by a "principle of commodification": whatever exists acquires its right to be because it is a potential commodity, something that can be transformed into a product to be marketed and sold. Such an economic model is dangerously beholden to a pervasive delusion, the tacit supposition that the economy functions in an inexhaustible, open-ended arena. To the contrary, the economy is a human field nested in the folds of the wider, more inclusive field of the biosphere. Because the biosphere is finite, fragile, and exhaustible, because it is delicate and vulnerable to damage, all economic activity occurring within its embrace must respect its finitude and fragility. The hubris of imagining the capacity for economic growth to be unbounded will eventually invite nemesis: the destruction of both the economy and the world community.

Our present economic model also presupposes that human

beings are quasi-mechanistic systems driven by the repetitive cycle of desire, acquisition, and gratification. Hence the market treats people as essentially consumers, aiming to provoke in them ever-new desires to acquire and enjoy the commodities it turns out by sucking in and transforming the biosphere. A life revolving around production and consumption means people are persistently beset by insatiable craving, restlessness, and a chronic sense of lack: the ills of modernity. When viewed against this background, the crisis of global warming can be read as an object lesson on the limits of the prevailing paradigm of market economics. It points to the need to develop an alternative economic vision grounded in a more respectful conception of human nature. In place of the old model of market capitalism, in which the driving force of the economy is the exchange of money for goods and services, we need an economy governed by the principle of sufficiency: one which reveres the intrinsic beauty and value of the natural world; gives priority to the full development of the human person; promotes a wide sense of human community; and addresses the broad dynamics of what constitutes human happiness.

In brief, the crisis of global warming compels us to look far beyond the immediate exigencies that a changing climate itself portends and assigns us the task of forging a new vision of human life, one that can integrate our endeavor to achieve a satisfactory standard of living for people all across the planet with the need to actualize our highest potentials while cherishing the biosphere that nurtures and sustains our lives.

But practical issues must be attended to first, as a matter of overwhelming urgency. Since, in earlier generations, the U.S. has been the beacon and leader of the democratic world, it is incumbent on this country to take the leadership role in reducing global warming. Before presuming to address others, we must ourselves set an example of ecological responsibility. This would require that we cut current levels of U.S. global warming pollution 25 percent by 2020 and 80 percent by 2050. To be effective, such controls must be mandatory and enforceable with heavy penalties for infringement. The federal government must make every effort to move America off oil by promoting fuel-efficient vehicles and establish low-carbon public transport. We must make a determined effort to generate electricity from clean, renewable sources like wind, solar, and geothermal power, which are established, zero-pollution technologies. With determined effort, and proper funding from the federal government, these technologies can be exponentially improved, and become practical replacements for the destructive, carbon fuels on which we have been relying.[5]

But while the U.S. should take the vanguard in resolving climate issues within its own borders, global warming is a problem to which many nations contribute and one that affects all peoples even in the most remote parts of the Earth. Thus nothing short of a global solution will work. What is needed, initially, is an international agreement that can tackle global warming across the entire industrializing world. This is the challenge that the world will face at the Copenhagen climate summit at the end of 2009, perhaps our last hope to redeem the planet from calamity. The agreement that emerges from this conference must recognize the responsibility of the industrialized world for our current environmental plight, and it should also be sensitive to the needs of developing nations to expand their economies. But such an agreement would have to be uncompromising in its insistence that everyone contributes, to the maximum degree, to the transition to a sustainable economy. There is no longer time for countries to seek their own advantage by wiggling their way out of firm commitments. Certainly we will have to experience pain; certainly we must be ready to place curbs on our self-indulgent lifestyles. But such sacrifices that we make to curb global warming are miniscule compared to those we would have to make if major climatic disasters become an everyday reality.

The agreement arrived at would have to focus first on "damage control," limiting the amount of damage inflicted on our planet by fossil fuels and environmentally destructive agricultural practice. Such damage will continue for decades until we can make a full transition to a green economy. The settlement must aim to keep emissions and global average temperature below the level at which runaway global warming would be triggered. To standardize emissions reductions, the U.S. has to collaborate and cooperate with developing countries like China, India, and Brazil and also with developed countries that have high per capita emissions such as Russia and Japan. If the U.S. is to play a leadership role in the campaign against global warming, it needs to offer a fresh vision that matches the full gravity of the problem. U.S. representatives must adopt a detached and objective view recognizing the claims of all nations equally and according each its place within a total solution, without seeking competitive advantages for their own economy. This could well be our final opportunity. We need an agreement that will allow both the developed and developing nations to cooperate in the crucial task of stabilizing the world's climate.[6]

Damage control would be only the first step, a major step, a necessary step, but by no means the last step. Along with imposing strong limitations on greenhouse gas emissions, the international community must collaborate to transform our present economic model so as to overcome the adversarial relationship between economic productivity and ecological sensitivity. Instead of competing in a relationship where one gains at the expense of the other, the economy and environment must join in a symbiotic bond—one where a thriving economy does not damage the environment, and ecological stability does not prevent people from achieving a comfortable, healthy, happy life. The new sustainable economic model must ensure that its benefits are widely distributed, so that no one need live in debasing poverty. Being centered around green technologies, it should be able to generate abundant energy without harming the beauty and integrity of the Earth.

The time has come to acknowledge that the role of fossil fuels in our economies and in our lives has been at best transitional. Their use has served to weld the world into an interconnected whole with a shared economy, culture, and institutions, but their role is coming to an end, and the threat they pose to our natural habitat demonstrates that the time has come to phase them out. This is the time for humankind to embark upon a new historical epoch, a post-carbon epoch. But we ourselves have to make the critical decisions, individually and collectively, that will determine our future destiny. We are now like the deity dwelling in the Judas tree. The golden goose has spoken to us time and again; we have failed to heed its warnings, and now the banyan is beginning to coil its roots around our trunk. We must hasten to uproot this banyan shoot before it is too late, so that we may live securely on this unique blue-green planet circling in its orbit around the sun.

Robert Aitken Roshi (b. 1917) is senior American Zen master and a Dharma heir of Yamada Koun Roshi, head of the Sanbo Kyodan in Kamakura, Japan. As a prisoner of war in 1941, he was introduced to Zen Buddhism by R.H. Blyth, and went on to practice extensively in Japan after the war. In 1974 he was given approval to teach, and in 1985 transmission as an independent master, by Yamada Roshi. The author of over ten books on Zen Buddhism, he founded the international Diamond Sangha network, and was one of the founders of the Buddhist Peace Fellowship.

Robert Aitken Roshi

Woe Unto Us!

Woe unto them that call evil good, and good evil; that
put darkness for light, and light for darkness; that
put bitter for sweet, and sweet for bitter!
Woe unto them that are wise in their own eyes, and
prudent in their own sight!
Woe unto them that are mighty to drink wine, and
men of strength to mingle strong drink:
Which justify the wicked for reward, and take away
the righteousness of the righteous!
Therefore as the fire devoureth the stubble, and the
flame consumeth the chaff, so their root shall be as
rottenness, and their blossom shall go up as dust.

—*Isaiah 5: 20–24*

According to Isaiah, the Lord shall wreak retribution on those who indulge in blameworthy conduct, but even without the intervention of the Lord, cause leads inexorably to effect. What goes around comes around:

Woe unto us, we who keep quiet about injustice,
murder, and torture!

> Woe unto us, we who stop at the pump as primordial ice falls into the sea!
> Woe unto us, we who bury the poisons of Lucifer that can't possibly stay buried!
> Woe unto us, we who conspire to upset the original balances that made all life possible!

Therefore flooding shall reduce all to rubble, and fires shall consume the chaff. Our homes shall collapse as babies die in the arms of their mothers, who then soon follow. Shakespeare himself shall vanish, together with Bach and Rembrandt. Kumarajiva shall vanish, together with Zhaozhou and Yunmen. The entire animal kingdom shall be gone. The entire bird kingdom shall be gone. Tall grasses and insects shall inherit the Earth!

HURRICANE KATRINA was among the first "extreme weather events" to affect collective consciousness about the potential devastation of global warming. On August 28th, 2005, driven by unusually high sea-surface temperatures, Katrina was a category 5 storm with winds of 175 mph. On August 29th, it made landfall as a category 3 hurricane with 125 mph winds and a 28-foot storm surge, becoming the most economically destructive storm in history (losses estimated at $125 billion) and causing the world's first mass movement of climate change refugees (a million people).

Joanna Macy, Ph.D., is a Buddhist scholar and eco-philosopher. A leading voice in movements for peace, justice, and deep ecology, she has interwoven scholarship with four decades of activism. She has created a ground-breaking framework and workshop methodology for personal and social change. Her books include *Despair and Personal Power in the Nuclear Age*, *Dharma and Development*, *Widening Circles*, and *World As Lover, World As Self*. Many thousands of people around the world have participated in her workshops, *The Work That Reconnects*, focussed on helping people transform despair and apathy in the face of overwhelming social and ecological crises into constructive, collaborative action.

Joanna Macy

On Being with Our World

To give full attention to the perils confronting our world invites an almost excruciating tension. It is the tension between seeing the enormity of the peril—such as climate chaos, mass extinctions, nuclear warfare—and seeing the inadequacy of our response to it. It takes courage to endure these tensions, yet endure them we must; for to be conscious enough to act responsibly requires being awake to the possibility of failure.

In my work in the environmental, peace, and justice movements, I share my conviction that spiritual practices can provide the moral strength to see things as they are. It is my experience that spiritual grounding, especially in the Buddhadharma, can keep us from shutting down or succumbing to wishful thinking. Practices that steady the mind and open the heart help us to be more present to our world.

Buddhist teachings bring me home to this beautiful, suffering world. They let me glimpse my non-separateness from it. In those moments I experience my inter-existence with all that is, and realize that there is, ultimately, no need to fear. That knowing has been reinforced by many of the people with whom I have worked, students and teachers and organizers from all walks of life and faith traditions.

Yet I also encounter people, including some on the Buddhist path, whose notions of spirituality hinder them from engaging with the world and realizing their power to effect change. Among the "spiritual traps" that cut the nerve of compassionate action, are these:

That the phenomenal world is an illusion. Impermanent and made of matter, it is less worthy than a realm of pure spirit. Its pain and its demands on us are less real than the pleasures or tranquility we can find in transcending them.

That suffering is a mistake. Pain we may feel in beholding the world derives from our own cravings and attachments. According to this view, freedom from suffering is attained by non-attachment to the fate of all beings, rather than non-attachment to matters of the ego.

That we create our world unilaterally by the power of our mind. Our subjective thoughts dictate the form things will take. Grief for the plight of the world is negative thinking. Confronting injustice and dangers simply creates more conflict and suffering.

And the corollary, that the world is already perfect when we view it spiritually. We feel then so peaceful that the world will become peaceful without our need to act.

We are free, of course, to seek escape from the suffering of our world, but the price for such comfort is high. It is my experience that the world itself has a role to play in our awakening. Its very brokenness and need call to us, summoning us to walk out of the prison of self-concern. And as we do that, as we venture into danger and uncertainty in service to life on Earth, discoveries await us.

These discoveries are symbolized by two hand gestures or mudras of the Buddha. On carvings and paintings we see them.

One is the *bhumi sparsha* or Earth-touching mudra that recalls the act that Gautama made when, at the outset of his vigil beneath the bodhi tree, he was challenged by Mara. Here the archetypal figure of Mara represents the fears and distractions that keep us from awakening to our unity with all life.

When Mara demands to know by what authority Gautama seeks the cessation of suffering, the one about to become the Buddha quietly touches the Earth. Instead of offering any personal credentials or curriculum vitae to show his worthiness to awaken for the sake of all beings, he touches the Earth, our ground of being. The central doctrine that he will teach, our dependent co-arising with all things, gives *us* that same authority. Our inseparability from all that is gives us power to act on behalf of all beings.

Stemming from that profound understanding comes the second gesture of the Buddha that calls me into life. It is the *abhaya* mudra. With right hand raised, palm forward, it means Fear Not. Don't be afraid. You will never be severed from the web of life, for that is what you are.

The gesture is similar to the traditional American Indian greeting I saw in movies as a child. I greet you and show you my open hand. See, I carry no weapons. Do not fear.

When the perils and tensions of this planet-time seem hard to bear, these two mudras with all they connote help me stay present to my world.

Joseph Goldstein is one of the first American Vipassana teachers and a co-founder with Jack Kornfield and Sharon Salzberg of the Insight Meditation Society (IMS) in Barre, Massachussets. He has practiced different forms of Buddhist meditation since 1967, under teachers from India, Burma, and Tibet. In 1998, he co-founded the IMS Forest Refuge for long-term personal retreats. He is resident guiding teacher at IMS, and leads Vipassana and Metta retreats worldwide. His books include *The Experience of Insight, Seeking the Heart of Wisdom* (with Jack Kornfield), *A Heart Full of Peace*, and *One Dharma*, which explores the creation of an integrated framework for the Theravada, Tibetan, and Zen traditions.

Joseph Goldstein

"Except as we have loved, All news arrives as from a distant land"

When I was invited to write a short essay for this book, my first thought was that I didn't have much to contribute on the subject of global warming. Although I am aware of the magnitude of the problem, perhaps like many others, I have not spent much time reflecting on it or seriously considering what I could do about it. It was this response that then piqued my interest. Why *hadn't* I spent time thinking about one of the major problems confronting our planet? Why had it slid to the backburner of my interests?

Two related teachings from quite different traditions began to shed light on these questions, light that illuminates other important issues in our lives as well. The first is a teaching from the great twelfth-century Korean Zen Master Chinul. His framework of teaching is *Sudden Awakening, Gradual Cultivation.*

> Although we have awakened to original nature, beginningless habit energies are extremely difficult to remove suddenly. Hindrances are formidable and habits are deeply ingrained.

> So how could you neglect gradual cultivation simply because of one moment of awakening? After awakening you must be constantly on your guard. If deluded thoughts suddenly appear, do not follow after them... Then and only then will your practice reach completion.

We have probably all had moments of what we might call a sudden awakening to the truth of global warming: reading different newspaper accounts, watching Al Gore's impactful film *An Inconvenient Truth*, times even of deriding those who don't believe it's happening—"How could *they* not believe the obvious scientific truth of it all?" Yet those moments can quickly pass, and the beginningless habit energies of forgetfulness, other desires, and basic ignorance re-surfaced once again.

Here is where Chinul's emphasis on gradual cultivation can be a template for our own awakening. We need to repeatedly remind ourselves of the situation and not settle for a generalized understanding that climate change is a problem. We need to be willing to make some effort to keep ourselves informed, over and over again, so that we don't fall back into deluded thinking: "How could you neglect gradual cultivation simply because of one moment of awakening?"

What might motivate us to make this effort? A powerful motivation for doing this is the feeling of compassion. In the Buddhist understanding, compassion arises when we're willing to come close to suffering, not as an abstraction, but in the reality of how lives are affected. What do people do when unusually strong and more frequent hurricanes devastate their homes and means of livelihood? How do people find food when traditional rain patterns are disrupted, when glaciers melt and rivers dry up, when island nations are submerged? Are we willing to open to these situations of suffering with an immediacy of feeling? The poet, Mary Oliver, expresses the challenge of this in her poem "Beyond the Snow Belt":

> "...except as we have loved, / All news arrives as from a distant land."

A second teaching that offers insight into the problem of rationalized disinterest is found in the words of Shantideva, an eighth century Indian adept. He wrote,

> We are like senseless children, who shrink from suffering but love its causes.

None of us desire suffering, whether it be the consequences of climate change or other painful circumstances of our lives, yet we are often addicted to the very causes of that suffering.

What is the way out of this unhelpful cycle? Ajahn Chah, the great Thai forest meditation master, said that there are two kinds of suffering: suffering that leads to more suffering, and suffering that leads to its end. If we can learn to understand the suffering and open to the reality of it, then instead of simply being overwhelmed by it, we can investigate its causes and begin to let them go. Here is where we can be a support for each other. Individually, we might feel that global problems are beyond our capacity to solve. What I have noticed, though, in the Insight Meditation Society community, is that if one or two people take the lead in making even small changes, it energizes the whole community. And if, for whatever reason, we don't feel ready to take a leadership role, it is helpful to acknowledge that and encourage those who feel inspired to do so. We can then be carried along in the slipstream of their energy, strengthening our own commitment in the process.

EARTH FROM SPACE: The first fully-lit images of the Earth floating like a "blue marble" in the darkness of space began a transformation of collective consciousness, demonstrating that we are a global community of people, animals, and plants sharing one planetary ecosystem and atmosphere. No single country has to live entirely within its own environmental means, but the world as a whole does. The current spikes in our greenhouse gas emissions, consumption, population, and species extinction are racking up an enormous ecological debt, whose payment is now coming due. *Courtesy of NASA*

Taigen Dan Leighton, Ph.D., is a Dharma successor in Suzuki Roshi's Soto Zen lineage, and guiding teacher of Ancient Dragon Zen Gate in Chicago. He also teaches at Loyola University, Chicago, and online at the Graduate Theological Union in Berkeley, California. He is the author of *Faces of Compassion: Classic Bodhisattva Archetypes and Their Modern Expression* and *Visions of Awakening Space and Time: Dogen and the Lotus Sutra*. He has translated numerous Zen texts including Hongzhi Zhengjue's *Cultivating the Empty Field*.

Taigen Dan Leighton

Now the Whole Planet Has Its Head on Fire

Collective Karma and Systemic Responses to Climate Disruption

For Buddhists to respond appropriately to the calamities that have only begun to befall us all from global climate disruption caused by human activities, we will need to rethink some common misunderstandings of karma that have prevailed in Buddhism. The teaching of karma has been frequently misused in Asian history to rationalize injustice and blame the victims of societal oppression. The popular version of this includes that people born into poverty or disability deserve their situation because of misdeeds in past lives. Such views have themselves caused great harm.

This ignorance also arises in the face of human atrocities, giving rise to absurd notions such as that people in the World Trade Towers on 9/11 deserved to die because their previous karma brought them to be there that day, or that victims of Hurricane Katrina deserved their fate because their past lives led them to live near failing levees. If we look fully at causes and conditions, the events of 9/11 cannot be separated from the unfortunate history of the Mid-East, and the complexities of the United States' and other nations' roles therein. The damage done by Hurricane Katrina cannot be separated from such factors as hurricane intensification due to climate change, the long history of American slavery and racism, and the complicated conditions through which the federal government in place at the time willfully disregarded warnings about the inadequacy of the levees.

Inflation of the effects of individual karma and lack of acknowledgment of collective karma ignore the basic Buddhist teachings of non-self and interconnectedness. *Anatman*, "non-self," is axiomatic to Buddhism; all merely individual selves are empty and lack inherent, substantial existence. True "Self" is the whole interconnected web of phenomena, depicted in the seventh century Chinese Huayan Buddhist school with the image of Indra's net, an intricate weave of connected interstices, each particular point of which completely reflects the totality of the vast network of being.[1,2] The application of Huayan thought to our environmental context has been clear to many modern Buddhists. Buddhist and environmental scholar Stephanie Kaza extends the metaphor of this jewelled net:

> If you tug on any one of the lines of the net—for example through loss of species or habitat—it affects all the other lines... If clouded jewels are cleared up (rivers cleaned, wetlands restored), life across the web is enhanced. Because the web of interdependence includes not only the actions of all beings but also their thoughts, the intention of the actor becomes a critical factor in determining what happens.[3]

Thus karma cannot be merely individual when our actions and intentions are so thoroughly connected with the whole environment. And our efforts toward collective, systemic responses have positive effects we cannot clearly measure or anticipate.

Going back to the thirteenth century, Japanese Zen master Dogen described the mutually beneficial impact of one person engaged in Buddha's meditation interacting with all phenomenal objects: grasses and trees, fences and walls, tiles and pebbles. Each element offers spiritual guidance for all others. He went so far as to say that with one person's sitting, "all space in the universe completely becomes enlightenment."[4] Clearly the karmic impact of the individual expressed in such teachings is not separate from a whole range of collective events. The dependent arising of each phenomenal event is due to a complex web of causes and conditions. Collective entities, such as a nation or a culture or a species, no less than individual human beings, have patterns of conditioned activities based on prior group actions. Multitudes of layers and levels of such communities are always involved.

The failure of the teaching of individual karma was recognized by the great leader of the Untouchables, Bhimrao Ambedkar, when he led the mass conversion to Buddhism of over three million Indian "Untouchables" starting in 1956. Ambedkar chose Buddhism for his fellow Untouchables after his careful study of world religions, but he rejected the Four Noble Truths. He felt that this teaching blamed individuals for suffering, and ignored "the heartless action of others and the systemic injustice of such social arrangements as the caste system. The idea of karma... would only accentuate the self-blame of the Untouchables instead of placing the blame on the caste system itself."[5] The second Noble Truth says that suffering has a cause, which is grasping based on desire, with resulting harmful karma. But this pattern of grasping and attachment is not necessarily, or even primarily, a simply individual matter. Other modern Asian Buddhists have also seen the necessity of responding collectively to sources of suffering. For example, the Sarvodaya Shramadana movement in Sri Lanka has benefited villagers and reduced suffering by seeing the need to inquire into collective needs in the villages, and mobilizing a communal response through appropriate work projects and ongoing communication.[6,7]

Responding coherently to global climate change requires our recognition of the reality of collective karma, and not seeing karma as merely individual. Individuals who practice recycling and modest use of resources contribute to helping the situation, but no matter how many individuals do so personally, the wider systemic causes also demand collective action to produce any significant effect. Without major development of sustainable mass energy sources, and governmental regulation of industrial and energy corporations to lessen their negative impact on the environment, the climate disruption will worsen, despite the personal practices of well-meaning individuals. Just as the causes for the situation are collective, a cooperative, systemic response is required, addressing the societal as well as individual karmic conditions for pollution and excessive carbon emissions. A Buddhist response now must involve study of systemic conditions for damage. In the case of climate change, this may include a wide range of practical studies, along with traditional Buddhist environmental perspectives.

Many current works may serve as resources for raising awareness and responding to systemic causes. Examples include Al Gore's film *An Inconvenient Truth*, with its vivid depictions of the scientific facts about climate change. Another perspective is offered by Naomi Klein's *The Shock Doctrine*, with its analysis of the workings of what she calls "disaster capitalism," in which corporate profiteers enrich themselves in the midst of disasters, which they sometimes encourage as well as exploit. She discusses how this applies to environmental as well as political crises, including the non-response to those endangered by Hurricane Katrina.[8] A rethinking of our current economic systems may be necessary for addressing global climate issues, which are not separate from the current global economic crisis. From the perspective of Buddhist "right view," no single ideological explanation by itself can encompass the total range of causes from many realms. Still, Buddhist practice confirms that awareness is transformative, both individually and collectively. Raising awareness, our own as well as that of all around us, will help make possible the larger changes that are needed.

As Al Gore among others has clarified, the situation of global climate change is urgent. Rapid systemic changes to use sustainable energy sources and stop emitting carbon are critical. Writings by the American Zen patriarch and poet Gary Snyder have illuminated the plight of our environment, and an approach to response. He clarifies the need for a wide perspective: "The larger view is one that can acknowledge the pain and beauty of this complexly interrelated world."[9] Snyder addresses the harm to the planet from "the highly organized societies and corporate economies of the world. Thousands of species of animals, and tens of thousands of species of plants, may become extinct in the next century. To nourish living beings we must not be content simply to have a virtuous diet." Collective, systemic response is vital.

Buddhist teaching, with its cosmological view of many arrays of buddha-fields throughout vast reaches of time, may also provide us a wider, useful temporal horizon now. Buddhist cosmology suggests beginningless and endless cycles of kalpas or ages, perhaps even translatable as many cycles of Big Bangs, past and future. Snyder has employed the standard Zen image of our heads on fire, the "great matter of life and death," used commonly as a challenge to individual Zen students. Now the whole planet has its head on fire. Going back to the sixties, Snyder has been saying that we need to act in the current situation as if our heads are on fire, but also we must simultaneously proceed as if we have all the time in the world.[10] Even amid an urgent crisis, it is most effective to act in a calm, deliberate manner. The systemic social changes that will help mitigate the worst effects of climate change will require persistent attention and response to be realized.

Returning to the individual level, Buddhist practice is excellent preparation for the inevitable damage likely to occur around us. The harm already triggered may at least be lessened by systemic responses, as well as by our own immediate response to the suffering involved. Meditative practice helps individuals to develop more calm and patience, a wider capacity to be helpful in the face of distress. In the midst of the immediacy of the next Hurricane Katrina, the specifics of which may likely not be apparent beforehand, the resource of our practice experience will allow each of us to be more skillful and flexible in responding to the suffering around us. And our individual response will be more helpful as we are willing to face our connection with the collective nature of the problem and its creation. We can both respond individually to crises, and work together to create societal responses to all the systemic factors that have brought trouble to this world.

Susan Murphy is an award-winning filmmaker and writer based in Sydney, Australia. She received Zen transmission from John Tarrant, a Dharma successor of Robert Aitken Roshi. She is the founding teacher of Zen Open Circle and a member of the Diamond Sangha Teachers' Circle. Her Zen teaching employs the koan tradition, but also includes artistic practice, dream-work, and exploring the resonance of the Dharma in Australia with Aboriginal spirituality. Her most recent book is *Upside-Down Zen*.

Susan Murphy

The Untellable Nonstory of Global Warming

Can We Really Be Allowing our Planet to Die?

My own response to global warming is still uncomfortably close to feeling frozen in the headlights of the juggernaut, a juggernaut that I am also part of. It seems to me we all are in some measure, almost helplessly, "living by damage" as Aboriginal elder Hobbles Danaiyarri put it, while the Dharma we convey is the deepest possible injunction against living carelessly. The contradiction feels increasingly powerful to me.

Climate change, the carefully neutral substitute ("Say, a change might be fun!") for the more pointed "climate damage" or "global warming," is unmistakably underway, with potentially catastrophic effects upon all current species adapted to viable climate niches, including us. Arguing the toss of exactly how much anthropogenic sources are contributing to this rapid change feels uncomfortably close to deferring facing an emergency, one that will demand extremely challenging, nearly unthinkable changes in how we think and act.

I think all people right now sense that the present abuse and despoiling of the life resources of the planet by our species are approaching intolerable and unsustainable levels. Why not begin with the issue most in our faces, which also happens to be the primary driver of this process—the likely effects of our massive over-consumption of carbon upon planetary heating; the demonstrable effects of that upon acidification of the oceans, pollution of the atmosphere, and accelerating loss of species; as well as the obvious speed at which it is at present driving population growth and propelling a mentality of "necessarily" exponential economic growth that seems entirely willing to turn a blind eye to the consequent destruction of people, animals, rivers, oceans, the air we breathe?

The whole world just watched an Olympic Games staged in a city shrouded in unbreathable air. This cannot go on. Up to now we've been like Auden's man on a runaway horse—asked "Where are you going?" we shout back, "I don't know, ask the horse!" When will we get round to asking the horse? And ourselves.

In one way, every trauma on such a scale, in which the secondary and subsequent "intergenerational" nature of the rippling effects of traumatization are so massive, is by nature a kind of untellable nonstory: almost impossible to get our heads around because the anxiety and shame it arouses is so strong it confiscates the issue from thought for much of the time. We simply haven't got our minds and tongues around it yet, we're too much inside its toils. And yet, or perhaps "and so," we have no choice but to try to wrestle it to painful consciousness and some ability to act effectively, offering whatever we can find in our means; we must certainly bring Dharma to bear as powerfully as we can on thinking into this and lifting it as skillfully as we can into the light of Dharma. What else is really possible?

Climate damage constitutes an act of harm toward all sentient beings on such a scale that it surely must rivet the urgent, critical attention of anyone who takes the Bodhisattva vows on a regular basis—or teaches Dharma. The climates and corresponding agricultural practices to which huge human populations are now minutely adapted, and to which we have geared vast expectations and massive infrastructure of human lives, are merely the provisional, semi-stable climate "settings" of the latest of a series of geologically brief interglacials—the one in which the critical move of human agriculture, and all that it has set in motion in terms of prevailing upon nature, was able to occur. But climate right now has become the most striking exemplum of "this too shall change," not just in daily fluctuation but progressive dramatic transformations across time. We seem as humans strangely able to live both as if there is no tomorrow, and as if we have a just right to demand that tomorrow shall be guaranteed to arrive. We surely must do whatever we can to disturb this terrifying sleep and wake ourselves up. That's the first move of any kind of response to the emergency. But it has to be a waking up to sobriety and coherence, not the arousing of panic or shame, which leads only to further dissociation.

Does it finally matter to what precise extent carbon dioxide emissions are the cause of the rapid loss of ice and glaciers, the increase in nighttime temperatures, rise in sea levels, onset of drought, and more frequent catastrophic storms, etc.? In any case, conserving limited fossil fuels as much as humanly possible and inconveniencing ourselves to do so, instead of falling over ourselves to release this vast reservoir of sequestered carbon as rapidly as we can into planetary circulation, surely makes its own strong sense. Isn't it an inherently good thing right now, possibly even to the point of actively paying its custodians to keep it in the ground?

The transformation wrought in the past 110 years by feeding the dense energy of oil into the economic bloodstream of change and "development" has been beyond anyone's dream or control; but now we are hooked, in just about every capacity—agriculture, medicine, clothing, transport, defense, housing, technology, furniture, and cosmetics are all intricately woven through with dependency on petroleum, and in grave danger of collapse if the fix is withdrawn, or even threatened with withdrawal. It seems anything, everything in fact, can be put at stake to ensure supply. And while we have almost no life(style) sustaining system that can now do without it, yet we are drilling and burning it as if there is no tomorrow—with a haste that implies we want to make sure there is none.

Wouldn't we—to the extent that a functioning "we" still remains in a time of mass individualism—want to be as conservative as possible toward the remaining stocks while doing all we can to propel and steer an urgent transition to a new energy economy, that must become as weaned and independent as possible from oil, coal, and gas? And to carefully conserve stocks of coal, for the production of longer-term needs like steel production, rather than for the momentary generation of power? As Australians, we live in a shameful, ostrich-like avoidance of acknowledging our truly crushing national carbon footprint, when you factor in the furious mining and export of our huge coal reserves for someone else to burn. The commodity boom, as long as it may last—that's all that matters to us.

How do we set about rigorously weighing up the relative harms between different courses of action and inaction? This opens a huge set of nested questions that will exercise every part of us and grow us up mightily, if we take it on, as an exercise in greater self-awareness. The shared peril of global warming is an overwhelming "moral opportunity," as Al Gore has said, an opportunity to restore ethical energy to a worldview almost religiously given over to rapacious self-interest. The first, salutary question here is what moral ground do we still actually possess to try to check the kind of ecologically damaging behavior in developing industrial economies that "we" in the post-industrial world have already profited by for generations, happily jeopardizing the living world as we leveraged ourselves to massive advantage? Meanwhile we scramble after Canadian tar-sands and the sudden possibility of Arctic Sea oil; any possibility at whatever cost. It seems we'd kill our mother and our grandmothers too, and even by implication our grandchildren's world as well, to keep up supply. Addiction is a too forgiving account for this behavior.

If we as a species were to be truly guided by the appropriate primary goal of removing the gross injustices of global poverty—rather than maintaining our own accustomed gravy train while trying to let on only as few new passengers as will keep it rolling nicely but not upset it for us—what would those energy choices and efforts toward innovation really look like? Might they not turn out to be the most creative and productive and economically energizing innovations of thought and practice possible for our time and for global well-being? Many would by their very nature involve a turn back toward local sourcing, local awareness, local community, more frugal mindset. Could the sheer exigencies of this moment be exactly the force needed to squeeze such valuable legs from the snake?

And since this is not the priority being presently set very much at any high level of decision-making, especially as all eyes turn to minimizing the impact of the credit "crunch," how do we begin to address the vast institutionalization of greed that every one of us now lives within, however uncomfortably or unwillingly, that sets this so low in the sights of the juggernaut of every "growth economy"? When you look to your own life, there's an immediate problem. How many facets and activities of your life can you actually find that are truly independent from "the system"? Well, meditation, along with making music, expressing love, jumping in the ocean, looking up at the sky, walking in the bush, and hugging your children—there's still a handful of precious things not yet invaded. What opens up from this?

Dharma is in one profound sense the Earth speaking to us directly, with our own noise no longer overwhelming the signal, and it is our response as natural and immediate as the perfect timing of every drop of rain falling onto the veranda beside me as I write. As teachers, our covenant is surely with the Earth and our Dharma has to rise to meet this occasion of planetary moral emergency. Do we have any choice but unflinchingly to face what is and try to respond with the intention of saving the many beings? Isn't the obvious first priority in an emergency to address, relieve, and forestall trauma with every means at our disposal that we are able to discover?

It's easy to sound a bit ratty and shrill, when the shape and vocabulary of the change so needed in such a dire moment is not yet fully in reach. This is one of the things that shuts us all up at exactly the wrong moment. In an old, wise fairy story, though, a woman who has had her hands chopped off by the Devil regains them in the sheer extremity of the moment that her child falls into a well. I take some heart from what this points to, about how to proceed.

In my heart I feel every human on the planet must surely be

raising a powerful lament right now, enough to deafen every world leader. Can we really be allowing our astonishing living planet to die, as a place that can continue to host even our own species? Perhaps it is still wholly an inward cry. But the relative planetary silence scares me deeply. Does it alarm you too? How are you handling it?

Matthieu Ricard (b. 1946), a Buddhist monk of Shechen Tennyi Dargyeling Monastery in Nepal, completed a doctoral thesis in molecular genetics at the Institut Pasteur in 1972, prior to moving to the Himalayas, where he became the close student and attendant of Dilgo Khyentse Rinpoche (d. 1991). His photographs of the spiritual masters, landscape, and people of the Himalayas have appeared in numerous books and magazines. He is the author and photographer of *Tibet, An Inner Journey* and *Monk Dancers of Tibet* and the photo-books *Buddhist Himalayas, Journey to Enlightenment,* and *Motionless Journey*. A dialogue with his father, French philosopher Jean-Francois Revel, *The Monk and the Philosopher,* became a bestseller in Europe. He is a board member of the Mind and Life Institute, and received the French National Order of Merit for his humanitarian work in the East. *Photograph courtesy of M. Stalens*

Matthieu Ricard

The Future Doesn't Hurt... Yet

Interdependence is a central Buddhist idea that leads to a profound understanding of the true nature of reality. Nothing in the universe exists in a purely autonomous way. Phenomena can only appear through mutual causation and relationship. The understanding of the laws of interdependence naturally leads to an awareness of universal responsibility, as is often pointed out by the Dalai Lama.

Since all beings are interrelated and all want to avoid suffering and achieve happiness, this understanding becomes the basis for altruism and compassion. This in turn naturally leads to the attitude and practice of non-violence toward human beings, animals, and the environment. The Buddhist idea of non-violence is not passive: it entails the passionate and compassionate courage to actively protect life and the environment. It nurtures a global vision of respect, care, and fulfillment.

Unchecked consumerism operates on the premise that others are only instruments to be used and that the environment is a commodity. This attitude fosters unhappiness, selfishness, and contempt. On the other hand, the Buddhist view that all sentient beings are endowed with buddha nature, and the universe in which they live is a buddha-field, shapes a culture of harmony and contentment.

The vast majority of Tibetans have never heard of global warming, although it is a well-known fact that the ice is not forming as thickly as before and the winter temperatures are getting warmer. In parts of the world where there is access to information, most of us are aware of the impeding danger of global warming and of the lack of serious measures taken by political authorities to address it. Even the "Stern Review on the Economics of Climate Change" that warned of the catastrophic economic impact of global warming had little impact on decision makers.[1] It is not as if more facts are needed; the evidence is striking enough.

People usually only consider changing their way of living when they are forced to do so by circumstances, not by rational and altruistic thinking. But in the case of climate change, once the dramatic events have occurred, and people become motivated to change things, it will be too late.

People react strongly to immediate danger but it is difficult for them to be emotionally moved by something that will happen in ten or twenty years. They will rarely be motivated to change on behalf of something for their future and that of the next generation. They imagine "Well, we'll deal with that when it comes." They resist the idea of giving up what they enjoy just for the sake of disastrous long-term effects. It is uncomfortable to drive a smaller car or be careful with water use or sunbathe judiciously. Their actions are based on not being inconvenienced *now*. The future doesn't hurt—*yet*.

In Nepal, for example, it is generally known that a magnitude eight earthquake could cause fifty thousand fatalities in a few minutes. On a daily basis, no one wants to plan with reference to an earthquake that happens on average every fifty years, even though it is now twenty years overdue. So, houses keep getting built that are not earthquake-proof, with no thought to future eventuality.

The only solution to the climate issue is for governments to adopt powerful new policies, even though they will not be popular at the moment. This requires a trans-national consensus and political will. For instance, the governments of most civilized nations have abolished the death penalty because it has been shown to be inefficient as a deterrent to crime, even though opinion polls show that the majority of people still are in favor of it.

When I attended the World Economic Forum in Davos in 2006, there was not much discussion about what might be done to prevent the Arctic from melting. In the last few years, however there has been a genuine increase of awareness about the causes and outcomes of global warming. There is now a substantial movement to recognize and take action on climate change. However it takes time for scientific knowledge to be implemented in public policies. It took thirty years from the time it was clear that smoking causes lung cancer to the point when political and legal action was taken to forbid smoking in public places in many countries. In some countries, such as China, chain smokers freely continue to poison everyone in a bus or a train, without the slightest restraint.

The Europeans are advancing with their renewable energy programs, but in the large Asian countries change is barely beginning and will require major shifts in policies and financial investments. It is difficult to expect poor truck drivers in Nepal to stop using old vehicles that emit clouds of black soot exhaust, since that would deprive them of their basic livelihood. In Europe people change their cars every five years. In Nepal, they keep them for twenty-five years because they lack funds to buy new ones. Who will give electric cars and efficient solar cookers to all these people? How are we to offer bio-gas to a billion people in India? Who is going to pay for all that?

The Chinese government is building a super-ecological island where everything will be zero-carbon, in an effort to show off their technology. Meanwhile, they do the opposite in the rest of the country—stimulating a car-buying frenzy, and allowing such unprecedented pollution of air and rivers that popular revolts are triggered in cities where acute environmental toxicity is harming people.

The Chinese government's approach to environmental issues is most often ineffective and chaotic. For example, the authorities' limited attempts at reforestation in Tibet are usually irrational. They plant trees on level land near rivers, whereby Tibetans engaged in agriculture are displaced. Mountainsides are not reforested, because it is more difficult to do so. The wrong species of trees are used, simply because it is easier. This does not halt the erosion of the slopes, and does deprive people of their crops. In the few remaining large forests, clear-cutting can continue. China might wake up to the implications of global warming and glacier retreat on the Tibetan plateau once these problems worsen.

The major constructive influence on environmental protection is now the European Community's sustainable energy policy. If America joins this initiative, it would instigate overall change. Large-scale adoption of wind power and other alternative sources of energies in the USA would have worldwide significance. Within ten years we could make substantial investments in renewable energy, and as time goes on, it will become less expensive. That is how technologies evolve: DVD burners cost $5,000 when they first came to market; now you can get them for $50. The oil billionaire Boone Pickens, a case in point, has put several billion dollars into wind power. Did he do it for the money? "Of course," he said, "the oil business is just mad. Renewable energy not only makes sense, but makes money as well." Even from the point of view of this Houston oilman, it makes sense. Such an individual can shift the perspective of other business people. If the USA begins to act, that could be the social tipping point for meaningful reduction of carbon emissions.

Let us keep hope.

Hozan Alan Senauke is vice-abbot of Berkeley Zen Center, California, where he lives with his family. He received Dharma transmission in the Soto Zen lineage of Suzuki Roshi from his teacher Sojun Mel Weitsman at Tassajara Zen Mountain Center in 1998. He is the founder of the Clear View Project, developing Buddhist-based resources for relief and social change. He is senior advisor to the Buddhist Peace Fellowship, an organization in the forefront of American engaged Buddhism. A poet and musician, he has studied and performed American traditional music for forty-six years.

Hozan Alan Senauke

The World Is What You Make It: A Zen View of Global Responsibility

We offer ourselves to ourselves, and we offer others to others. The causal relation of giving (*dana*) has a power that pervades the heavens above and the human world below... Entrusting flowers to the wind and entrusting birds to the season may also be the meritorious action of giving.

—*Dogen*, The Bodhisattva's Four Methods of Giving

Looking around this troubled planet, we see fierce and unusual weather—last year's Cylone Nargis in Burma, 2005's Hurricane Katrina in the U.S., unprecedented storms flooding low-lying areas of Asia, and tornadoes sweeping across the American midwest. The slow but measurable warming of oceans is breaking up age-old glaciers in Greenland, Iceland, Canada, and the Arctic regions. Inland, drought is steadily sucking away lakes and wetlands, leaving whole populations to fight over diminishing water resources. Where I live, in California, we now have a year-round season of wildfires. The future bodes more of the same.

Sixty-five years ago, human technology "developed" to the point where nuclear weapons gave us the means to destroy Earthly civilization in one blow. After an initial horrendous experiment with atomic bombs in Hiroshima and Nagasaki, we have—so far—taken the road of nuclear deterrence, refraining from further use of such weapons. But our heedless pattern of consumption, our seemingly unquenchable appetite for fossil fuels, has led us to destroy an environment that has sheltered us for age after age. In geological time, the present environmental destruction flowing from human activity over less than a hundred years—the massive generation of carbon dioxide and resultant global warming—is as swift as a nuclear bomb, and potentially as destructive to life.

From our narrow human perspective on time, global warming seems to have come on slowly. But this is hardly the case. A public awareness that we, people, are causing global warming is something new, hardly discussed even by scientists until the 1970s. Today, though, it seems to be settled science. But despite high-minded government rhetoric and self-congratulatory hoopla about the supposed "greening" of multinational corporations, our addiction to fossil fuels has hardly waned. In fact, the "developing" world is increasingly folded into a technological and consumerist global economy, swelling the world's appetite for oil, coal, natural gas, and deforested agricultural land.

Let's change gears here... When the Buddha awakened, so long ago, his original teachings turned on the wheel of *paticca samuppada* or, in English, dependent origination. The ins and outs of dependent origination are highly complex but its core principle has been clearly expressed in early Buddhist texts.

When this is, that is.
From the arising of this comes the arising of that.
When this isn't, that isn't.
From the cessation of this comes the cessation of that.

The Buddha's early teachings pointed to dependent origination as the driving wheel of birth, death, and rebirth. In traditional Buddhist terms, this is seen as playing out in each individual life after life. Toward the middle of this wheel we find six realms of existence into which beings are born. The realm of hungry ghosts or *pretas* features ghastly beings with huge, swollen bellies and long, pencil-thin necks. They are insatiably hungry, but unable to swallow the food they crave. Pretas are driven by illusory appetites and desires that can never be fulfilled or satisfied. We all know people like this, who are never satisfied, never have enough material riches and things.

From the vantage point of engaged Buddhism, with its structural and systemic view of dependent origination, present-day corporate globalization is exactly an expression of how the Buddha's wheel of dependent origination gives rise to whole nations, peoples, and cultures born and reborn in the realm of hungry ghosts. I live in such a nation. It is called the United States.

I dwell in the United States, because that is where I live, and, hopefully, that is where my engaged Buddhist practice and political work are most likely to have an impact. The U.S. uses around 25 percent of the world's annual energy resources to support roughly 5 percent of the world's population. Our 200,000,000 cars and trucks, most of them less than fuel efficient, account for much of the world's carbon dioxide emissions. In 2004, U.S. vehicles generated the equivalent of 314,000,000 metric tons of carbon. Imagine a coal train 55,000 miles long, stretching around the globe twice! Half of our electricity comes from aging coal-fired power plants, which release, along with vast amounts of carbon dioxide, large quantities of sulphur dioxide and mercury, causing acid rain and toxicity. So-called "clean coal technologies" are still far in the future. In the agricultural sector, chemical fertilizers and waste products from ruminant animals account for 10 percent of U.S. greenhouse gases, far surpassing the impact of agriculture in other parts of the world. As the saying goes, we live "high on the hog." And people around the world pay the price for our lifestyle.

Consider Cyclone Nargis, which devastated Burma in May of 2008. A U.S.–based meteorologist called Nargis, "one of those once-in-every-500-years kinds of thing." But the Centre for Science and Environment, a respected Indian environmental monitoring group, sees Cyclone Nargis as an effect of global warming, a sign of things to come. Sunita Narain of CSE said "the victims of these cyclones are climate change victims, and their plight should remind the rich world that it is doing too little to contain its greenhouse gas emissions." She added that large-scale polluters bear responsibility for what is happening in Burma.

If this is true, then our connection to Burma, and other nations where unforeseen storms, floods, and drought are appearing with greater frequency, is a matter of dependent origination—"because there is this, that arises"—of cause and effect. Could it be that the cyclone's devastation arises from our addiction to fossil fuels, which causes global warming, rising sea levels, and new weather patterns? Global warming is just one effect of our addiction to fossil fuels. Our hunger for Burma's oil and natural gas (along with the energy appetite of developing nations like China, India, and Thailand) is precisely what provides Burma's brutal military dictatorship the economic leverage to stay in power. I do not wish to live by stealing the natural resources of impoverished people half a world away nor by causing environmental destruction that endangers everyone, myself included. If our national habits of consumption contribute to world's hardships, then what are we to do?

The Zen Buddhist tradition, as passed down to us in the West from China and Japan, is, like Tibetan Buddhism, an expression of Bodhisattva practice. Bodhisattvas vow to save beings from suffering, reaching out, warm hand to warm hand. We live this vow by following the Bodhisattva's moral precepts of thought, speech, and action—no killing, stealing, lying, intoxication, and so on. These are precepts of relationship, of giving and receiving. As Zen Master Dogen says: "We offer ourselves to ourselves, and we offer others to others." These principles can guide our actions as individuals, communities, and nations.

Here are some practices for healing the world. These are not solutions, but tools for looking at ourselves and what we do. Each of them is nearly impossible to realize. I take this phrase—"nearly impossible"—as a challenge to do all that I can.

- The Bodhisattva precepts boil down to one essential principle: not to live at the expense of other beings. This is simple to say, and very difficult to do.

- Each of us must take complete responsibility for the world, as if the world's fate depended on our words and actions. Whether we know it or not, it does.

- An old Zen teaching says, "Not knowing is most intimate." But this is different from knowing nothing or willfully closing one's eyes. Considering the suffering of myself and others, naturally I study everything I can find. It is when I study and perhaps feel I am developing some mastery or understanding, that I know that my understanding is incomplete and will always be so. That incompleteness is "not knowing." To accept that about oneself and to press on is to be intimate with the world.

- Act mindfully and correctly, irrespective of results. Do things because they are the proper things to do. It may seem as if one's own modest efforts at conservation have no impact, but recall the Jataka tale in which a parrot carried water, drop by drop, to save his forest from spreading fires. His single-minded and seemingly hopeless dedication inspired a god's tears, which quenched the flames.

- Thinking globally, acting locally is good, but limited. One must also think locally and act globally. That means simultaneously working to curb consumption at home, at work, in one's town, and pushing our elected representatives to enact legislation and policies that have impact on a national and international level. Our national moral authority flows from a willingness to make personal sacrifices.

- "Entrusting flowers to the wind and entrusting birds to the season..." When I recognize that my life and everything in it has been freely given to me, how can I deny this gift to all other beings, and to the planet itself? Take only what one needs and allow all things to be free and fully themselves.

I have no idea if such practices and policies will prove successful. Despite the efforts of our best minds and most powerful computers, we don't know what the future will bring.

But the world is what you make it...

Lin Jensen is the author of *Pavement,* which chronicled his experiences as a protester for peace, *Together Under One Roof,* and *Bad Dog!* He is the founding teacher and senior teacher emeritus of the Chico Zen Sangha, in Chico, California, where he lives with his wife.

Lin Jensen

The Rising Temperature of Planet Earth

Global warming is inextricably linked to patterns of consumption, a direct result of what and how much we humans buy. Patterns of excessive economic consumption are pretty much the same throughout the industrialized world. Here in the United States, for example, the conventional economic viewpoint holds that when the index of consumer confidence is up, that's good. When it's down, that's bad. The more goods people buy, the better the prospects for the economy. Traders on Wall Street respond by buying up shares, the Dow industrial averages rise, and most people feel positive about their economic prospects. The fewer goods people buy, the worse the prospects for the economy. Traders sell off shares, the Dow industrial averages fall, and people feel pretty dismal about economic prospects. On the whole, people in affluent countries expect to buy what they want when they want it. Not satisfied with the simple necessities of food, shelter, clothing, and medicine, they want things like a another pair of shoes to complement the half dozen they've already got, a membership at the sports club, a new stereo for the car, the leather jacket they've been coveting in the window of

Saks Fifth Avenue. The fact that people already in possession of a great many things are seldom satisfied with what they've got is worth reflecting on.

What's clear to me is that the invariable offspring of consumerism is more consumerism, that wanting engenders wanting, and that no lasting satisfaction is likely to result from getting what I want. Like greed, of which it is an expression, wanting is self-perpetuating until opposed by self-restraint. Twenty-six hundred years ago in India, the Buddha observed that the hunger to possess was a source of dissatisfaction that, among the other forms of suffering, was the suffering of not getting what you want. But he went on to explain that it was the wanting itself that constituted the suffering, the craving for things rather than the simple lack of them. We'd get along quite contentedly without the latest high-resolution TV, if we didn't give rise to thoughts of its possession. But we do give rise to thoughts of possession and are encouraged to do so by every conceivable marketing device available. And from this circumstance arises a chronic dissatisfaction just as the Buddha said it would. We become restless to spend and own, and once conditioned to seek satisfaction in that manner we find no end to it.

The cost of this compulsive consumption to the Earth and to its inhabitants is tremendous. It is this craving to possess with its attendant production and waste that has brought us the fearful crisis of global warming. A thermometer is placed under the tongue to measure the presence or absence of fever, which if found is symptomatic of a pathological infection. The rising temperature of planet Earth measures just such a fever, diagnostic of an epidemic of unrestrained human greed.

I can't credit my own relative restraint in this matter, because I've never earned more than a modest salary. I don't know what harm I might have done had I had the money to do it with. What I can say is that I've never really wanted a lot of things that others find indispensable. As a youth I didn't have to cope with the endless enticements to purchase that today's youngsters deal with. I grew up at a time when credit cards were unheard of and people worked fulltime just to supply the essentials. Virtually all that my family owned was visible from the back porch: a kitchen garden with a few fruit trees, a milk cow, a half dozen pigs, and some chickens running about the yard were the sole source of our living. We were content to have this much, and thought very little of what we didn't have, perhaps because what we didn't have was mostly out of reach. I don't ever remember feeling deprived, and I don't consider that impression as simply a child's naïve view of things. I honestly recall that we were perfectly happy. Why do I now indulge in memories of a period of United States history occurring over sixty years ago? I do so because it was a time when an industrialized nation of considerable wealth was forced to curtail its consumption of material goods, discovering in the process that a pleasing and agreeable life was consistent with such constraints. Sometimes unexpected benefits were derived from these "lean" years. In England virtually no meat was available during the war, and the rate of heart attack plunged to an all-time low, rewarding a mild dietary austerity with increased cardiovascular health.

Now as an adult, I've adopted a Buddhist economic heritage characterized by modesty and restraint. The early Sangha that gathered around the Buddha was a model of voluntary simplicity. The term "patched-robe monk" literally describes the only clothing a monk possessed, a robe patched together from bits of discarded cloth and sewn by hand. Aside from his robe and sandals, a monk's belongings typically consisted solely in what he could carry on his person—a bowl for alms and few utensils. Among householders, however, the Buddha actually encouraged a moderate accumulation of wealth, providing it was attained by ethical means in an amount not exceeding one's needs. What he condemned was hoarding wealth at the expense of others, a constraint that alone limits consumption to a degree that is utterly unknown in the consumer-driven economy that characterizes most developed industrial nations today.

If I want to apply the principles of Buddhist economics to my own patterns of consumption, I must learn to distinguish between *need* and *want*. The *needs* of any one person, household, or township are finite, while *wants* are without limit. *Wants* reside in the mind, a product of thought, while *needs* are of the body, consisting of such reasonable necessities as food, clothing, shelter, and medicine. A simple analogy makes the distinction more tangible: *wanting* to eat is eating when you *feel* like eating; *needing* to eat is eating when you're hungry. It's a distinction upon which the survival of the Earth's delicately balanced ecosystem relies.

In a little mountain valley at the foot of California's Mt. Shasta lies Shasta Abbey, a Soto Zen monastery. To enter the Abbey is to enter an economic refuge where the standard of living isn't measured by the capacity to consume, where needs are distinguished from wants, and an unspoken moratorium on unnecessary consumption is in effect. Working one day in the Abbey kitchen, I was given a crate of lettuce to make ready for inclusion in a salad. It was discarded lettuce thrown out by a market in the nearby village of Mt. Shasta, as too far gone for public sale. It was pitiful, a wilted mess that I couldn't imagine anyone would be asked to eat. But the monk in charge of the kitchen patiently explained to me how to tear away a few of the more blemished outside leaves of the lettuce and put the rest to soak in a large tub to re-hydrate. When the lettuce had soaked awhile, I was surprised to see how revived it appeared to be. The leaves had filled out and begun to look like something I might even consider eating. With a few other greens, tomatoes, grated carrot, and parsley from the Abbey garden added, we all sat in the dining hall and ate a perfectly delicious summer salad. It was a great satisfaction to know that we could feed ourselves on what others throw away.

As I write these words on December 1, 2008, the nation and the world has fallen upon a massive global economic recession. The major world stock markets have lost over 50 percent of their value, banks are failing in record numbers, foreclosures on homes are unprecedented, the ranks of the unemployed swell, and the worldwide gross national product shrinks. Admittedly it's a mess. But instead of trying to buy our way out of it, could we just this once do what nature does so well and try saving our way out of it? In times such as these, the distinction between luxury and necessity reasserts itself, and I'm forced to set aside what I merely want for what I actually need. I recall how I once stood in line outside Vandermast Clothiers, clutching two precious World War II–era coupons allotted me, one for the purchase of a pair of Levi jeans and another for a pair of school shoes. We all kept our place in line, knowing that supplies might run out before our turn came. I can't help but compare that orderly line of earlier times with the frenzied mob of shoppers who, on the day after Thanksgiving 2008, gathered outside a Long Island, New York, Walmart to participate in the nation's newly instituted shopping spree, and who in their lust for bargains trampled to death a temporary employee who happened to get in the way.

We won't rescue ourselves from the mess we're making of things by discounting prices. Getting a top-of-the-line television for just $599.95 won't do a thing to prevent the melting of the Arctic ice and the subsequent starvation of the polar bear who depends upon the ice as a way to make a living. In the distress and failure of the polar bear, as of many other species, we humans witness our own distress and failure. In our haste to have and to own, we have taken whatever we've wanted from the Earth and returned as payment an atmosphere polluted by an excess of greenhouse gases. As the planet warms, we humans are literally purchasing our own demise.

We must change our ways or accept the ruin of ourselves and of the Earth upon which we depend. Can we now form among ourselves a modest and sustainable economic community dedicated to lives of voluntary simplification? Can we put aside individual wants so that the needs of all can be met? Can we once again learn to wait our turn in line?

Part V
Solutions
John and Diane Stanley

Clarity, Acceptance, Altruism: Beyond the Climate of Denial

The ever-present possibility of deceit is a crucial dimension of all human relationships, even the most central: our relationship with our own selves...

Lying is obliged by its very nature to cover its traces, for in order to lie effectively we must lie about lying. This poses a problem for anyone attempting to prove the ubiquity of deception. Although it is all around us, deception is strangely elusive.

—*David Livingstone Smith*[1]

Thousands of climatologists around the world have been researching global warming for three decades. There is no "scientific debate" about the existence or causes of our climate crisis—although there is intensive ongoing investigation of how dangerous, rapid, and consequential its impacts will be. However, society has reacted to the scientific facts largely through collective denial. This inability to acknowledge our true situation is a psychological defense mechanism: a mechanism of self-deception. Complex manipulative strategies always oil the wheels of society, and so it should not surprise us that collective denial has been encouraged and sustained by sophisticated efforts of social deception.

The game-changing Fourth Report of the Intergovernmental Panel on Climate Change (2007) represented the international consensus of scientists and government representatives—some of the latter being advocates for national fossil fuel industries. It was conservative but its conclusion was unambiguous: global warming is caused by human activity. The IPCC was deservedly awarded the 2007 Nobel Peace Prize. Meanwhile, something unprecedented was happening in the Arctic—a massive, unprecedented loss of summer sea ice, many decades ahead of the IPCC's projection. The concern of biologist Tim Flannery was abundantly verified:

> The pronouncements of the IPCC do not represent mainstream science, nor even good science, but lowest common denominator science—and of course even that is delivered at glacial speed... If the IPCC says something, you had better believe it—and then allow for the likelihood that things are far worse than it says they are.[2]

Politicians and corporations mostly now claim to accept that the cause of climate change is human activity, but continue to ignore fundamental opportunities for corrective action while carbon emissions continue to increase. We are accelerating toward catastrophic climate tipping points, while our global society remains fundamentally confused on the issue of its own survival—a kind of collective psychopathology.

In 1989, the Global Climate Coalition was formed by fifty fossil fuel and allied corporations. Its agenda was to cast as much "skepticism" as money could buy on the existence of global warming. These corporations spent hundreds of millions of dollars on public relations, advertising, and political donations, perpetrating intellectual, psychological, and environmental damage, and depriving science and representative democracy of any real influence over a crucial decade. When the scientific evidence reached a critical mass, a number of corporations withdrew, leaving Exxon Mobil, General Motors, and Chevron as unrepentant sponsors in 2000. The stage was set for political takeover of the American government itself by a coalition of forces including big oil, the military-industrial complex, and radical Christian fundamentalist groups for whom both evolutionary biology and climate change were heretical beliefs.[3]

The Global Climate Coalition was replaced by new "skeptic" organizations, backed by right-wing think tanks or the oil and coal industries. Many resort to blatant fraud to bewilder the public: the "Leipzig Declaration," for example, purported to be signed by dozens of senior academic scientists. Most "signatories" were non-scientists and it also included the names of scientists who had never seen it. "Climate skeptic" tactics follow a playbook established by tobacco corporations, who maintained huge profits through knowingly misleading the public for two decades after smoking had been shown to cause lung cancer. In the case of global warming, however, the price is not being paid by those who choose to smoke tobacco, but by all present and future life on Earth.

From 1995 to 2005, the number of scientific research papers on global warming published in international journals was 928, and in all cases there was no doubt as to the cause of global warming. In the same period, 636 articles appeared in the mainstream press, and over 50 percent of those cast doubt on the cause of warming.[4] This tragic disconnect between science and society was abetted at the highest levels of the U.S. government—for eight critical years between 2000 and 2008.

Science is rarely simple enough for non-scientists to fully understand, so the opportunity exists for spurious "experts" to bamboozle and deceive. Their false journalistic arguments are not subject to the rigorous peer-review process required for publication in scientific journals, and resemble a malevolent political smear campaign. Any attempt at a strategic solution to the climate crisis must begin with an accurate appraisal of the world as it is. It becomes necessary, therefore, to refute and debunk the pseudoscience of "climate skepticism."

Some "skeptics" have asserted that a natural phase of increasing solar radiation could be warming the Earth, just as a natural decrease of solar radiation cooled it in the "Little Ice Age" of the seventeenth century. The truth is that since the Industrial Revolution, the sun has been cooling while the Earth's climate has been warming. Solar changes can in no way explain the dramatic rise in temperature over the past few decades.[5] A second "skeptic" argument asserted that particulate air pollution is welcome, since it reflects sunlight and provides a cooling effect. This partial truth is at best of temporary duration. It was put forward to distract people from the overwhelming greenhouse effect of carbon gas emissions.

Another "skeptic" argument asserted that a decrease of atmospheric particulates from volcanoes, which temporarily reflects solar radiation back to space, was responsible for global warming. To the contrary, the world's volcanoes have been very active since the 1950s, the period when climate warming took off dramatically. A fourth claim was that climate warming results from a regular force of nature that manifests when the Earth "wobbles" on its axis. Changes brought about in this way take over one hundred thousand years. How then could they explain the extremely rapid warming effects since 1950? Lastly, another argument attributed climate change to the warming effects of El Niño, which last only a few years and operate on selected geographical areas. To the contrary, what the world's climate is going through is global and cumulative—dramatically so over the last six decades.

The world's most powerful supercomputers at the U.K. Hadley Climate Centre calculate millions of information signals per second—comparing and dynamically integrating solar radiation, moisture, air, heat, interactions with ocean and land, volcanoes, and greenhouse gases. Among every combination of factors examined, the only one that parallels the drastic global warming of the past six decades is the increasing atmospheric concentration of greenhouse gases—above all, carbon dioxide generated by human activity. During the Carboniferous Period (from 300 to 360 million years ago) photosynthesis withdrew that carbon dioxide from the ancient atmosphere, but over the last century it has been released again by our fossil fuel economy.

> Human activity is to blame for the rise in temperature over recent decades, and will be responsible for more changes in the future… If anybody tells you differently, they either have a vested interest in ignoring the scientific arguments or they are fools.
>
> —*Gabrielle Walker and Sir David King*[5]

Advertising and PR tactics do not invoke pseudo-scientific reasoning, but take another approach to mass deception. Apparently innocuous "opinion pieces" by certain journalists, ex-politicians, or captains of industry are placed in popular newspapers, employing dismissive cognitive framing to undermine alternative energy. Ongoing campaigns of scientific-looking adverts emphasize the altruism of big oil and conflate the word "energy" with fossil fuels. Advertising campaigns, lavishly-funded by the coal industry throughout the U.S. election year (and onward), employ the oxymoron "Clean Coal" as a cognitive frame to befuddle the public. We are told we can trust a non-existent, unproven technology—carbon capture and sequestration—so that construction of new coal-fired power plants may continue, as emissions rise and renewable energy is delayed. The most polluting fossil fuel of all is thereby marketed as clean enough to save our climate.

Biased information is also provided to governments and the media through institutionalized forms of deception. The Energy Watch Group recently analyzed how the International Energy Agency (IEA), the body that advises most major governments on energy policy, has consistently obstructed a global switch to renewable energy systems—because of its ties to the oil, gas, and nuclear sectors. IEA reports have repeatedly underestimated the amount of electricity available from wind, displaying "ignorance and contempt" toward it, while promoting oil, coal, and nuclear as "irreplaceable" technologies. The inverse is actually the case. On current trends, world wind capacity will reach 7,500GW by 2025, and fossil fuel power stations could be phased out completely by 2037.[6]

Denial and deceit lead to systemic stress and despair in a society. How are we to begin to heal this confusion and suffering? The factors that can assist us include introspection, clarity, and a proactive response. The simplest form of proactive response involves listing, precisely and realistically, the actions needed to change the situation. The second option is to change the way one sees the problem; for example, we can see global warming as a crisis or an opportunity. Thirdly, after thorough investigation, we can stop resisting the situation and accept the reality as it is. From this acceptance, clarity and resources can emerge. We can then take the best course of action. Thich Nhat Hanh states:

> If we don't know how to stop our over-consumption, then the death of our civilization will surely come more quickly. We can slow this process by stopping and being mindful, but the only way to do this is to accept the eventual death of this civilization, just as we accept the death of our own physical form. Acceptance is made possible when we know that deep down our true nature is the nature of no-birth and no-death.[7]

Along with the growth of consumerism and carbon gas emissions, health professionals have observed a striking rise in the rate of psychological depression—in America, more than tenfold in the last fifty years.[8] We clearly need to change our focus from over-consumption, waste, and reckless "growth" to the well-being, morale, community, and safe climate people really need. Through urgent action, individual and collective, we can achieve these "authentic goods." It is an act of healing to commit to a sustainable way of life, to organize ourselves and help organize our communities to protect the Earth. The Dalai Lama often states that the first beneficiary of altruism is oneself. As Thich Nhat Hanh points out, we already have a full range of renewable energy technologies, but we do not take advantage of them because of collective despair, anger, division, and discrimination.[7]

Our society needs to acknowledge that technology does not work without brotherhood, sisterhood, understanding, and compassion:

> Meditation is not an escape. It is the courage to look at reality with mindfulness and concentration. Meditation is essential for our survival, our peace, our protection. In fact, it is our misperceptions and wrong views that are at the base of our suffering. Throwing away wrong views is the most important, most urgent thing for us to do.
>
> —*Thich Nhat Hanh*[7]

DRY LANDSCAPE ZEN GARDEN, KYOTO, JAPAN: Kyoto, with its classic Buddhist temples and gardens, seemed an auspicious location to negotiate a climate protection accord, but the treaty of 1997 did little to slow massively increasing greenhouse gas emissions. As early as 1958, D. T. Suzuki presciently observed in *Zen and Japanese Culture*: "Zen purposes to respect Nature, to love Nature, to live its own life; Zen recognizes that our Nature is one with objective Nature, not in the mathematical sense but in the sense that Nature lives in us and we in Nature... Instead of raising the so-called standard of living, will it not be far, far better to elevate the equality of living? At no time in history has such a truism been more in need of being loudly declared, than in these days of greed, jealousy and iniquity. We followers of Zen ought to stand strongly for the asceticism it teaches."[9]

ONE OF THE DRAGONS from "The Nine Dragons" handscroll by Chen Rong, 1244 CE. In many East Asian cultures, dragons are revered as representative of the primal forces of nature, religion, and the universe. They are associated with wisdom and longevity. The dragon is therefore an appropriate symbol for renewable energy.

A Renewable Future

> The time has surely come when we must speak out as Buddhists, with firm views of harmony as the Tao. I suggest that it is also time for us to take ourselves in hand... This would be engaged Buddhism where the Sangha is not merely parallel to the forms of conventional society and not merely metaphysical in its universality... This greater Sangha is, moreover, not merely Buddhist. It is possible to identify an eclectic religious revolution that is already underway, one to which we can lend our energies.
>
> —*Robert Aitken Roshi*[1]

Buddhist meditation is based on refined attention, the introspective examination of inner experience. Sophisticated brain-scanning techniques have revealed neuro-protective effects (e.g. on the aging process in the brain) as well as neuro-plastic effects (distinctive anatomical changes to the brain in response to intentional meditative exercise). Experienced meditators demonstrate awareness and choice at deep levels of inner attention—characterized by alpha, theta, or gamma brainwave activity—together with integrated right and left brain hemisphere functions. Large clinical studies confirm that meditation training also has important therapeutic applications for ordinary patients suffering from anxiety, stress, and depression. We might describe it as an inner renewable energy system.

Outer renewable energy systems are beneficial from a Buddhist perspective because they are limitless, benign, and harmonious with nature—characteristics of the Dharma itself. They are essentially free, after an initial investment. In contrast to carbon fuels, the price of electricity from renewable sources falls rather than rises with time. Renewable energy systems are an indispensable external solution to the main causative factor of global warming. Furthermore, everything we need to make use of the world's abundant carbon-free energy—wind, solar, hydro, tidal, and geothermal systems—is available right now.

Society's leaders have responded to the urgent scientific warnings about global warming by clinging to "business as usual" in the face of more and greater ecological evidence. In this institutionalized denial, the primary obstacle to progress is the wealthy but failing fossil fuel economy itself. We have begun to experience systematic failures throughout global capitalism, due to accumulated fraud, debt, and price volatility predicted for Peak Oil.[2,3] The outcome is economic depression. This will last as long as it takes for us to eliminate the waste of energy for profit, and to fully de-carbonize our energy systems.

Experience in Germany, as well as major studies,[4,5] show that renewable energy systems create quality jobs that cannot be shipped overseas, displace carbon pollution, and provide a secure, sustainable electricity supply. There is no doubt that renewable energy will have to replace fossil fuel this century. The question is whether we will adopt it in time to save ourselves from societal collapse, and save our planet from climate collapse.

> At the heart of a genuine climate-stabilizing initiative is a detailed plan to cut carbon dioxide emissions 80 percent by 2020 in order to hold the future temperature rise to a minimum. This initiative has three major components—raising energy efficiency, developing renewable sources of energy, and expanding the Earth's tree cover... We are in a race between tipping points in natural and political systems. Which will come first? Can we mobilize the political will to phase out coal-fired power plants before the melting of the Greenland ice sheet becomes irreversible? Can we halt deforestation in the Amazon basin before it so weakens the forest that it becomes vulnerable to fire and is destroyed? Can we act fast enough to save the Himalayan glaciers that feed the rivers of Asia?[6]

Sweden is leading the way to an oil-free economy by 2020. Germany, Denmark, Portugal, and Spain have radically increased energy autonomy by rolling out wind power and other renewable energy systems. All over the world, countries are now following suit. Al Gore has boldly and persuasively argued that five major initiatives would produce 100 percent of American electricity needs from renewable sources within 10 years. The first is construction of concentrated solar thermo-electric plants in the Southwestern deserts, wind farms in the corridor stretching from Texas to the Dakotas, and advanced geothermal plants in hot spots. The second is a unified national smart grid for the transport of renewable electricity from rural places to the cities where it will mostly be used. This will comprise new high-voltage, low-loss underground lines designed with "smart" features to eliminate inefficiency and reduce customer bills.

Thirdly, the failing American car industry should convert quickly to manufacture plug-in hybrids that can run on renewable electricity. Fourthly, nationwide retrofitting of buildings with better insulation, energy-efficient windows, and lighting would eliminate most of the 40 percent of U.S. carbon emissions that originate from them—while saving money for homeowners and businesses. Fifthly, the U.S. should put a price on carbon, and lead the world to replace Kyoto with a more effective treaty to cap and reduce global carbon gas emissions, and sharply reduce deforestation.

To implement this energy revolution on a planetary scale, internationally agreed policies and treaties are essential. They will include large-scale government investment—money that would otherwise be wasted on oil subsidies and oil wars. Innovative new laws can specify financial incentives and environmental regulations that set forth strict efficiency standards for vehicles, appliances, and buildings. Compensated access to the electricity grid can be guaranteed for all renewable energy developments. A carbon tax with 100 percent dividend could return funds to households based on how much they reduce their carbon footprints. Someone who his or her carbon footprint more than average would actually make money. Fossil fuels can be taxed at the well, mine, or port of entry to create the most efficient behavior.

The deployment of renewable energy on the requisite scale will require industry to be re-tooled. It is encouraging to recall that the very effective American domestic war effort of the 1940s was accomplished without computing, automation, or robotics. Scaling such a program up worldwide would be far easier now, and it would abolish the root causes of climate breakdown and energy insecurity through a single policy shift. Its scientific and socio-political foundations have been described in the pioneering works of Lester Brown[6] and Herman Scheer.[4]

> We have entered the Century of the Environment, in which the immediate future is usefully conceived as a bottleneck. Science and technology, combined with a lack of self-understanding and a Palaeolithic obstinacy, brought us to where we are today. Now science and technology, combined with foresight and moral courage, must see us through the bottleneck and out.[7]

THE WIND GOD by Ogata Korin (Japanese, Edo Period, eighteenth century). Wind power is abundant, cheap, clean, inexhaustible, widely distributed, and climate-benign. No other energy source has these six characteristics.

Five Transformative Powers

Wind Power

Wind Power is abundant, cheap, clean, inexhaustible, widely distributed, and climate-benign. No other energy source has these six characteristics.

If wind farms were constructed with today's large turbines, the U.S. could generate three times the amount of electricity now produced from coal and nuclear plants.[1] The largest onshore wind farm in Europe, located on twenty-eight square miles of moorland, will generate 320MW of electricity and power the city of Glasgow. Denmark obtains over 20 percent of its electricity from wind and has a goal of 75 percent by 2025. Scotland plans to obtain 50 percent within a decade.

Wind is intermittent, yet generating 25 percent of electricity from it does not require special back-up provisions. Wind intermittency can be predicted, managed, and mitigated in a variety of ways, such as the well-planned spatial distribution and integration of wind farms. Current barriers largely reflect the social, political, and technological inertia of the traditional electricity generation system. There is now a significant new energy-storage technology that eliminates intermittency through wind turbines that generate compressed air.[2] This is stored in an underground reservoir, and released to provide electricity-on-demand to the grid, for use at peak times.

Wind turbines can be sited offshore, where winds are strong and continuous. A 1GW (1,000MW) offshore wind farm, the "London Array" in the Thames estuary, will power 25 percent of Greater London's homes. The U.K. government plans 33GW of offshore wind power by 2020; enough to supply all domestic electricity demand in Britain.

The United States already has 24,000MW of grid-connected wind-generating capacity (equivalent to twenty-four coal-fired power plants) and a huge 225,000MW of planned wind farms waiting for access to transmission lines. It has forty plants manufacturing wind power components and many more under construction. Investment in wind farms creates four times the number of jobs created by coal-fired power plants.[1]

Solar-thermoelectric Power

This suite of technologies is also called Concentrating Solar Power. They concentrate the sun's heat to generate steam to drive a turbine or engine, converting sunlight to electricity with an efficiency of 15 percent. They overcome the intermittency of sunlight by heating molten salts like sodium and potassium nitrate up to 600°C (1112°F). The power plant can run into the night, providing base-load electricity to the grid even in the absence of sunlight. It can be scaled up to thousands of megawatts capacity. This is a powerful choice for sunny, arid, or desert lands in the southern USA, the Mediterranean, India, China, Africa, etc.

One advanced solar thermoelectric technology system, consisting of mirrors arranged in a concave plate, focuses sunlight onto a Stirling external combustion engine. It converts sunlight to electricity with an efficiency of 25 percent. This technology is being used for world's biggest solar-thermoelectric power plant (1,800MW), under construction in California.

Until recently there was only one operational solar-thermoelectric plant in the U.S. There are now eighteen commercial-scale plants under development (fifteen in California, two in Florida, and one in Arizona). This labor-intensive renewable energy technology, with its sharply falling cost curve,[1] is about to become a major player in U.S. and European energy economies.

Solar Photovoltaic Power

The first photovoltaic (PV) cell containing semiconducting material was created by Bell Labs in 1954. When hit by a photon of light, these materials liberate an electron that is guided into a circuit by a conductor. The electron leaves a "hole" to be filled from the other end of the circuit—creating an electric current. Early silicon cells converted light to electricity with 6 percent efficiency. Silicon cells will soon reach their limiting conversion efficiency of 30 percent, but other materials could achieve as much as 70 percent efficiency.

Germany has installed PV panels generating as much electricity as five large coal-fired power stations. Much sunnier countries—Spain, Portugal, and Italy—have followed this lead. In 2008, a leading California utility contracted with two companies to build 800MW of solar PV generating capacity. The peak output will equal that of a nuclear reactor.

For solar PV to be widely deployed, the cost of cells must be reduced. Accordingly, thin-film cells have been developed. A mixture of semiconductor compounds, amounting to a small percent of the material in regular cells, is sprayed onto a flexible base. Economies of scale for thin-film cells could make solar PV a major "grid-competitive" power source worldwide in only a few years.

Tidal Power

Twice a day the Atlantic Ocean pours 14 billion tons of seawater through the Bay of Fundy off Nova Scotia, Canada—said to be equivalent to the combined flow of every river on Earth. Nearly a quarter century ago, the provincial utility began a modest tidal power pilot project in the bay. It is still operating, producing 20MW of electricity every day, one of three professionally-run tidal power operations in the world.

Tidal power can now provide much more significant inputs to the electricity grid. In the U.K., a proposed ten-mile tidal barrage across the Severn estuary exemplifies the technology. The Severn Barrage would contain over 200 turbines (each 40MW). Arrays of sluices would let the tide in and then close, forcing water out through turbines after the tide has retreated some distance beyond the barrage. The Barrage would generate 17 million kilowatt hours of electricity annually, 6 percent of total U.K. demand. It would displace 18 million tons of coal, or 3 nuclear reactors. It would also provide flood protection to the coastal areas of the Severn Estuary.

The Pentland Firth is a stretch of water about twenty miles wide between the Scottish coast and the Orkney Islands. Independent reports show that it contains six of the top ten tidal energy sites in U.K. waters—and is one of the best tidal resources available anywhere in the world. Tidal power in the Pentland Firth could supply at least 8 percent of total U.K. electricity consumption.

Enhanced Geothermal Power

Geothermal energy is the heat emanating from the planet's interior. It is inexhaustible, limitless, and carbon-neutral. Geothermal power production requires heat, water, and fractured rock. Since heat is available anywhere under the surface of the Earth, water and fractured rock are what is sought in a "natural" geothermal site. For twenty years, the U.S. had a single commercial-scale geothermal power plant, in California. As of 2008, there are 96 projects in western states, the largest with a generating capacity of 350MW. This marks the emergence of a major new source of electricity.

The exciting new energy technology of Enhanced Geothermal Systems (EGS) artificially creates permeability in hot rock and introduces water to extract the heat. It is being actively developed in France, Germany, Switzerland, and Australia. A recent review by the Massachusetts Institute of Technology[3] showed that EGS is not limited to existing geothermal hot spots, but can be applied essentially anywhere. Since advances in oil drilling technology now permit drilling as much as six miles down into solid rock, water can be injected down one well, passed through crevices in the hot rock, and the steam extracted through another well. Properly supported government research programs would rapidly and dramatically reduce its cost.

EGS plants are not decades away: operational plants already exist at Soultz in France (1.5MW) and Landau in Germany (2.5MW). At Cooper Basin in Australia, a 500MW power plant will come online in stages by 2011. Developing EGS will provide expanded power production both in the near future, and even more in the long run. EGS is a far more highly developed technology than the merely hypothetical idea of carbon capture and sequestration. The latter is a completely unproven, and certainly expensive, concept for promoting so-called "clean coal." Indeed, EGS could produce 100,000MW of electrical power in the U.S. by 2050, displacing two hundred large coal-burning plants.[3] This comparison should be of critical importance to decision makers.

THE DISH-STIRLING ENGINE SYSTEM: One of the most efficient solar-thermoelectric technologies, this uses concave mirrors that track the sun and focus a large area of sunlight into a beam of concentrated light, for use as a powerful heat source. At the focal point is a Stirling engine, a closed-cycle regenerative heat engine with a permanently gaseous working fluid. This quiet, highly efficient engine converts heat to mechanical power (and thus to electricity) by alternately compressing and expanding a fixed quantity of gas at different temperatures.

LONDON AT NIGHT, from the International Space Station. *Courtesy of NASA*

The End of Energy Waste

One of the major factors driving global warming is easy and profitable to correct: our waste of energy. Due to technological advances, it now costs one-third as much to save energy through efficiency programs as it does to generate the same amount by building a new power plant. A major scientific report to the U.S. Congress recently stated:

> Energy end-use efficiency could reduce carbon emissions by approximately 60%, while renewable energy could easily make up the rest of a 70% emissions cut.[1]

Eighty percent of society's energy now comes from fossil fuels. Two-thirds of it is lost during conversion to forms used in human activities. The significance of this becomes even clearer when we examine the appearance of our major urban areas at night from space. So the fastest, cheapest, most effective way to reduce fossil carbon emissions is to avoid as many losses as possible. It could eliminate more than half our carbon emissions without affecting the gross domestic product.[2]

Up to 40 percent of carbon emissions come from buildings. Whether newly constructed or retrofitted to low-energy demand standards, their annual energy demand could be reduced six-fold. Very large energy savings are also available from energy-efficient appliances.

Lighting uses 25 percent of all electrical power generated in modern societies. Conventional incandescent bulbs waste 90 percent of the electricity they consume—they produce more heat than light. Compact fluorescent bulbs have superseded them in efficiency and are becoming mandatory in many countries. White light emitting diodes (LEDs) are the ultra-efficient light source of the future.

Goodbye to the Internal Combustion Engine

The Plug-in Hybrid Electric Vehicle

Plug-in HybridVehicles (PHEVs) contain extra batteries that can be recharged by connecting to an electric power source.They incorporate three energy sources: a petrol or diesel-fuelled engine, the regenerative energy from braking or excess engine power, and electricity from the power grid. In all-electric mode, driving costs are one quarter of those of a petrol-fuelled vehicle.

Remarkable liquid-fuel economy (over 100mpg) is a feature of PHEVs. Replacing the national vehicle fleet with PHEVs would reduce transport emissions and oil imports by 70 percent. No fossil fuel at all is used during the all-electric range, if batteries are charged from renewable energy sources like wind. A PHEV can be recharged at home using cheap-rate electricity.

In the proposed vehicle fleet system called "vehicle-to-grid," stationary PHEVs act as a storage system for wind-generated electricity. This overcomes the issue of wind intermittency and makes wind power the major fuel source for road transportation.[1]

The Compressed Air Technology Car

The zero-emissions "Air Car" uses compressed air technology (CAT).[2] It dispenses with the internal combustion engine altogether.

The Air Car's engine works by controlling the movement of four pistons and a single crankshaft. There is an electric moto-alternator on board that compresses air, recharges the battery, and acts as an electric brake. The vehicle uses a lightweight aluminium frame and carries carbon-fiber tanks to store compressed air. Fuel-air can either be compressed at home with its on-board compressor, or refilled in a few minutes at stations with air compressors. And if the external compressor is powered by renewable electricity, so will be the vehicle.

Generating high pressure is exactly why the internal combustion engine burns hydrocarbon fuel with air. The CAT engine extracts the energy stored in compressed air to drive the wheels. Compressed air at 290 times atmospheric pressure is like energy stored in a battery that you charge and draw from later.

The first Air Cars designed for urban markets have a range of 125 miles and a top speed of 70 mph. The parent company has produced an urban taxi, and a CAT/petrol hybrid with a 2000km (about 1250 miles) range. Air Cars entered European and Australian markets in 2008. The vehicle is cheap, and its running costs are 80 percent less than comparable petrol cars.[2]

Tradable Energy Quotas

We should be taxing what we burn, not what we earn.

—*Al Gore*

Individuals are responsible for 40 percent of carbon emissions. David Fleming has developed a realistic scheme of tradable personal carbon allowances.[1] Its object is to address the linked issues of global warming and hydrocarbon energy limitation—and to do so across the whole of society. Climate chaos is advancing rapidly, and Peak Oil is imminent. Creative carbon-rationing offers a practical solution for both. Fleming's Tradable Energy Quotas (TEQs; pronounced "tex") concept is a "cap and trade" scheme. It separates companies and government from individuals. It creates incentives to stop wasting energy and creatively make do with less.

Any country introducing TEQs first establishes its carbon budget—tons of emissions permitted yearly. The budget is set five years in advance by a politically independent energy policy committee. In the first year, the budget matches current emissions. Thereafter it falls yearly. Everyone knows in advance how much less carbon energy they will have to spend over the next twenty years. The budget is shared between people, companies, and government. People get their ration free—companies and public bodies must pay for theirs.

Your TEQ would come as units (one unit would be equivalent to one kilogram of carbon gas released). When you buy petrol, gas, or electricity from a coal-fired power plant, you surrender some of it electronically by debit card. If you consume less by year's end, you can sell the excess through electronic trading. If you consume more, you can buy more. Every year, the available pot of excess rations diminishes. The scheme provides an orderly, planned transition to a post-carbon economy. It offers incentives for business to develop, deploy, and market carbon-neutral and energy-efficiency solutions.

Renewable energy is zero-rated under TEQs, so it becomes cheaper than fossil fuels. Individuals get limits and incentives to participate in the transition to a viable future for all. The scheme is fair and redistributive, since everyone in the country starts each year with the same quota. We all become stakeholders in the problem and in the solution, rather than everything being left as the exclusive responsibility of government and big corporations. TEQs would empower individuals, and diminish collective feelings of helplessness about the climate emergency.[1,2]

Drawing Down Carbon with Biochar

Terra Preta (black earth) is a two-thousand-year-old man-made soil with enhanced fertility, due to high levels of soil organic matter and nutrients. Terra Preta soils occur in the Brazilian Amazon, Ecuador, and Peru in areas of about fifty acres—in the midst of surrounding infertile land. Cash crops like papaya and mango grow three times faster on terra preta.

These Amazonian black soils were created by pre-Colombian agriculturists and sequester twice the carbon of other soils. Charred, rather than burned, wood is stable in soils. It does not rapidly degrade, but becomes a fungal matrix and a water filter. As water flows through, it leaves nitrogen, potassium, and other nutrients trapped in the matrix. Fungi make them available to plants.

Now soil scientists are burning biomass at a high temperature in the absence of oxygen. This process produces a charcoal called "biochar" and a biofuel (oil or gas). The charcoal is a unique soil-restoration additive. The biofuel is burned to generate heat for the process itself, or electricity. The biochar process removes more carbon dioxide from the atmosphere than it releases. Growing plants have removed carbon dioxide from the air—then their charcoal sequesters it in the soil for thousands of years. This is an affordable technology that can be applied at village or industrial scale. It could draw down and safely store billions of tons of carbon in soils.

Biochar reduces the emission from soil of the potent greenhouse gas nitrous oxide. Adding biochar triples biomass yield—in the tropics, this increases the number of growing seasons per year. Nutrient-poor tropical soils resulting from slash-and-burn techniques support just a few agricultural crops before they are exhausted. Biochar would grant them "permanent fertility" and alleviate pressure for deforestation.

Biochar allows agricultural crop residues and other types of biomass now classified as "waste" to become biofuel sources. This would benefit rural economies worldwide. If the U.S. adopted a cap-and-trade program for carbon gas emissions, farmers in the Midwest could double their income by using the leaves, stalks, and cobs that remain after harvest to fuel biochar pyrolysis.[1]

Reducing the Carbon Footprint of the Meat Industry

The livestock sector of agriculture is a major contributor to global warming. To elucidate this, the UN Food and Agriculture Organization (FAO) studied the meat production chain. They found that livestock produce a larger share of greenhouse gas emissions than transport does.[1] Pasture and feed crops account for 9 percent of carbon emissions. A third of methane emissions come from ruminant livestock. Animal manure contributes 65 percent of nitrous oxide emissions.

Livestock now make up a fifth of all terrestrial animal biomass. So the sheer quantity of animals raised for human consumption has come to threaten biodiversity. The Worldwide Fund for Nature states that a third of the Earth's ecological regions are endangered by livestock production. Directly, or through the growing of soybeans for animal feed, it is the main driver of deforestation in the Amazon basin.

To mitigate global warming, the FAO makes two recommendations. Firstly, livestock and feed crop production should be spatially intensified, in order to reduce deforestation. Secondly, animal nutrition and manure management should be improved to reduce methane and nitrous oxide emissions. Consumers can exert the most effective political pressure on this industry.[1]

Worldwide, 627 million tons of grain are devoted to livestock feed. Americans eat 25 times more meat per capita than Indians, and this increases their risk of cardiovascular disease and common cancers. The second expert report of the World Cancer Research Fund, *Food, Nutrition, Physical Activity, and the Prevention of Cancer: A Global Perspective* (2007) suggested limiting red meat consumption to eighteen ounces or less per week.[2] Meat eaters could mitigate global warming, support biodiversity, and improve their own health by replacing much (or all) of their meat consumption with plant protein—by "eating lower down the food chain."

> The UN Food and Agriculture Organization has estimated that direct emissions from meat production account for about 18% of the world's total greenhouse gas emissions. So I want to highlight the fact that among options for mitigating climate change, changing diets is something one should consider.
>
> —*Dr. Rajendra Pachauri, Chairman, Intergovernmental Panel on Climate Change*

Ending Deforestation

At first I thought I was fighting to save rubber trees, then I thought I was fighting to save the Amazon rainforest. Now I realize I am fighting for humanity.

—*Chico Mendes, Brazilian Environmentalist*

Of all forms of habitat destruction, the most consequential is deforestation. Most species of plants and animals on Earth live in rainforests. Rainforests constitute an essential cooling band around the planet's equator. Twice as much carbon is stored in these trees as in the whole of the Earth's atmosphere. Poor countries continue to slash and burn these irreplaceable forests at huge rates: an area equivalent to the U.K. is destroyed every year. Carbon emissions from five years of this deforestation exceed that from all aviation from its beginning till about 2025 (based on current numbers of flights).

Since up to a quarter of carbon emissions come from deforestation, the fastest, cheapest way to reduce them is to protect forests—in Brazil, Indonesia, Congo, etc. This does not necessitate developing new technology, but political will, enforcement, and incentives to ensure that forests are more valuable when left standing. The protection of standing forests was absurdly left outside the Kyoto Protocol and its carbon markets. The 2007

IPCC report points to its inclusion in a new climate treaty as the best hope to avoid a catastrophic climate breakdown.

The REDD initiative (Reducing Emissions from Deforestation in Developing Countries) is a pay-and-preserve scheme for forests, launched in 2005 by the UN Framework Convention on Climate Change. Countries are paid to prevent deforestation, while funding comes from industrialized countries seeking to meet their emissions commitments under Kyoto. Policymakers and environmentalists value REDD because it fights climate change at low cost, improves living standards for poor people, and safeguards biodiversity.[1] It was the major positive proposal agreed at the 2007 Bali Climate Conference.

Reforesting the Earth

Under a tree the great sage Buddha was born.
Under a tree he overcame passion
And obtained enlightenment.
Under two trees he passed into Nirvana.
Indeed, the Buddha held trees in great esteem.

—*Dalai Lama XIV*[1]

Trees fundamentally beautify our planet and are important symbols for all religions. They provide very many protective functions and services. Those include home and shelter for mammals, birds, insects, and other species—and shade. Trees prevent desertification, stabilize soil, and conserve water. They control avalanches and floods. They provide essential oils, gums, resins, syrups, and vital medicines. They yield fruit, nuts, timber, paper, and fodder. Trees carry out the crucial unseen work of carbon sequestration. In one year, a typical tree inhales twelve kilograms of carbon dioxide and exhales enough oxygen for a family for a year.

Planting a tree is something everyone can do—from school children to old people, in both rural and urban environments and in any nation. To plant a tree is empowering and effective. The major strategies proposed to draw down carbon and secure a safe-climate future include reforestation of the planet on an unprecedented scale.[2,3]

> When we plant trees, we plant the seeds of peace and seeds of hope.
>
> —*Wangari Maathai, Nobel Peace Laureate 2004*

The United Nations Environment Programme has begun a worldwide Plant for the Planet: Billion Tree Campaign.[4] It originally committed to one billion trees each year. The response from individuals, schools, businesses, and organizations around the globe has been so positive that the goal is now seven billion trees by the end of 2009. More than 165 countries participate. The program encompasses anything from a village woman planting a single tree to a large corporation planting thousands of trees. One can also sponsor the planting of a grove of trees, through award-winning organizations.[5]

> The symbolism—and the substantive significance—of planting a tree has universal power in every culture and every society on Earth, and it is a way for individual men, women, and children to participate in creating solutions for the environmental crisis.
>
> —*Al Gore, Nobel Peace Laureate 2007*[6]

> We should also take a lead in the protection of forests and rainforests. Their destruction contributes greatly to climate breakdown, while their preservation cools the Earth and ensures its biodiversity. Indeed, we should be part of a global effort to plant many more trees and forests.
>
> —*Gyalwang Karmapa XVII*[7]

Part VI
The Bells of Mindfulness
Thich Nhat Hanh

Thich Nhat Hanh (b. 1926) is an expatriate Vietnamese Zen Buddhist monk, teacher, author, poet, and peace activist. During the Vietnam War, he founded the School of Youth for Social Services in Saigon, a grassroots relief organization that rebuilt bombed villages, set up schools and medical centers, and resettled families left homeless. He traveled to the U.S. to urge the government to withdraw from Vietnam, and led the Buddhist delegation to the Paris Peace Talks. It was during this period that Martin Luther King, Jr., nominated him for the Nobel Peace Prize. He skillfully adapted the core practice of mindfulness to Western sensibilities, creating the Order of Interbeing in 1966, establishing monastic and practice centers around the world and coining the famous term "Engaged Buddhism." His more than one hundred books include *Being Peace, The Sun My Heart, For A Future To Be Possible,* and a remarkable collection of poems, *Please Call Me By My True Names.* In 1991, he was awarded the Courage of Conscience award.

The Bells of Mindfulness

The bells of mindfulness are calling out to us, trying
to wake us up,
reminding us to look deeply at our impact on the
planet.

The bells of mindfulness are sounding. All over the Earth, we are experiencing floods, droughts, and massive wildfires. Sea ice is melting in the Arctic and hurricanes and heat waves are killing thousands. The forests are fast disappearing, the deserts are growing, species are becoming extinct every day, and yet we continue to consume, ignoring the ringing bells.

All of us know that our beautiful green planet is in danger. Our way of walking on the Earth has a great influence on animals and plants. Yet we act as if our daily lives have nothing to do with the condition of the world. We are like sleepwalkers, not knowing what we are doing or where we are heading. Whether we can wake up or not depends on whether we can walk mindfully on our Mother Earth. The future of all life, including our own, depends on our mindful steps. We have to hear the bells of mindfulness that are sounding all across our planet. We have to start learning how to live in a way that a future will be possible for our children and our grandchildren.

I have sat with the Buddha for a long time and consulted him about the issue of global warming, and the teaching of the Buddha is very clear. If we continue to live as we have been living, consuming without a thought of the future, destroying our forests and emitting dangerous amounts of carbon dioxide, then devastating climate change is inevitable. Much of our ecosystem will be destroyed. Sea levels will rise and coastal cities will be inundated, forcing hundreds of millions of refugees from their homes, creating wars and outbreaks of infectious disease.

We need a kind of collective awakening. There are among us men and women who are awakened, but it's not enough; most people are still sleeping. We have constructed a system we can't control. It imposes itself on us, and we become its slaves and victims. For most of us who want to have a house, a car, a refrigerator, a television, and so on, we must sacrifice our time and our lives in exchange. We are constantly under the pressure of time. In former times, we could afford three hours to drink one cup of tea, enjoying the company of our friends in a serene and spiritual atmosphere. We could organize a party to celebrate the blossoming of one orchid in our garden. But today we can no longer afford these things. We say that time is money. We have created a society in which the rich become richer and the poor become poorer, and in which we are so caught up in our own immediate problems that we cannot afford to be aware of what is going on with the rest of the human family or our planet Earth. In my mind I see a group of chickens in a cage disputing over a few seeds of grain, unaware that in a few hours they will all be killed.

People in China, India, Vietnam, and other developing countries are still dreaming the "American dream," as if that dream were the ultimate goal of mankind—everyone has to have a car, a bank account, a cell phone, a television set of their own. In twenty-five years the population of China will be 1.5 billion people, and if each of them wants to drive their own private car, China will need 99 million barrels of oil every day. But oil production today is only 84 million barrels per day. So the American dream is not possible for the people of China, India, or Vietnam. The American dream is no longer even possible for the Americans. We can't continue to live like this. It's not a sustainable economy.

We have to have another dream: the dream of brotherhood and sisterhood, of loving-kindness and compassion. That dream is possible right here and how. We have the Dharma, we have the means, and we have enough wisdom to be able to live this dream. Mindfulness is at the heart of awakening, of enlightenment. We practice breathing to be able to be here in the present moment so that we can recognize what is happening in us and around us. If what's happening inside us is despair, we have to recognize that and act right away. We may not want to confront that mental formation, but it's a reality, and we have to recognize it in order to transform it.

We don't have to sink into despair about global warming; we can act. If we just sign a petition and forget about it, it won't help much. Urgent action must be taken at the individual and the collective levels. We all have a great desire to be able to live in peace and to have environmental sustainability. What most of us don't yet have are concrete ways of making our commitment to sustainable living a reality in our daily lives. We haven't organized ourselves. We can't only blame our governments and corporations for the chemicals that pollute our drinking water, for the violence in our neighborhoods, for the wars that destroy so many lives. It's time for each of us to wake up and take action in our own lives.

We witness violence, corruption, and destruction all around us. We all know that the laws we have in place aren't strong enough to control the superstition, cruelty, and abuses of power that we see daily. Only faith and determination can keep us from falling into deep despair.

Buddhism is the strongest form of humanism we have. It can help us learn to live with responsibility, compassion, and loving-kindness. Every Buddhist practitioner should be a protector of the environment. We have the power to decide the destiny of our planet. If we awaken to our true situation, there will be a change in our collective consciousness. We have to do something to wake people up. We have to help the Buddha to wake up the people who are living in a dream.

Afterword

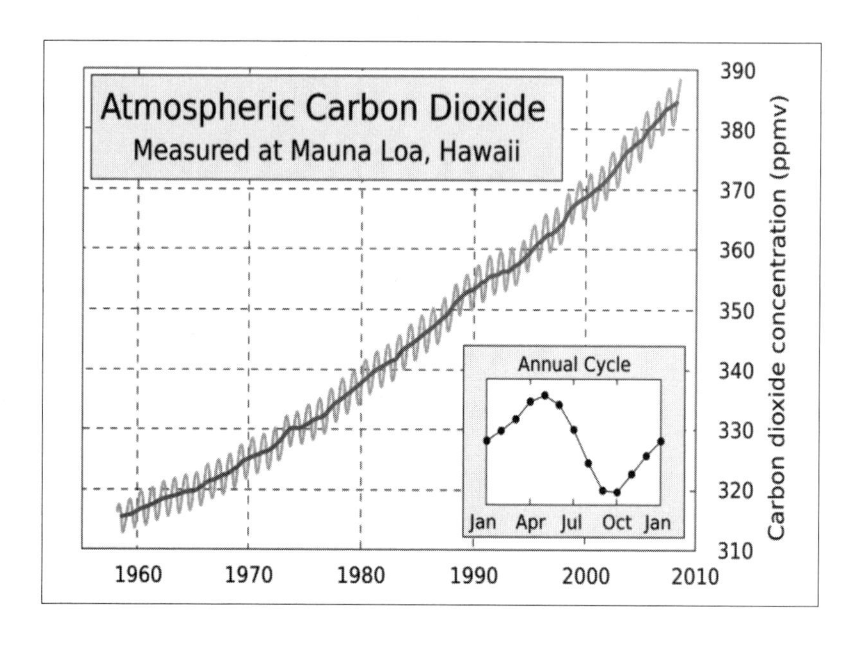

AT MANY SITES around the world, such as Mauna Loa in Hawaii, scientists have measured man-made increases in the greenhouse gas carbon dioxide, the main driver of global warming. Atmospheric carbon dioxide fluctuates annually, because more is "drawn down" during summer by large Northern Hemisphere forests. The annual cycle, shown in the inset figure, appears as "saw-teeth" behind the yearly average rise.

The pre-industrial atmospheric concentration of carbon dioxide was 280 ppm (parts per million). The current level is 387 ppm—the highest for 650,000 years, long before the modern human species appeared. It is still increasing, by 2 ppm per year. Where should humanity aim? The safe-climate target for atmospheric carbon dioxide is 350 ppm, the level that avoids the possibility of runaway warming and maintains the planet we know. *Courtesy of Robert Rodhe, Global Warming Art Project*

The Dalai Lama

Endorsement of a Safe Level for Atmospheric Carbon Dioxide

In his closing speech to the international climate talks in Poland in December 2008, Al Gore stated that former targets for fighting global warming had been rendered obsolete by new findings, and that 350 parts per million of atmospheric carbon dioxide was the new standard for which the world should aim. His remarks drew the longest applause of the conference. The 350 target accepts that we are challenged not only to reduce carbon gas emissions, but to actively remove huge quantities of fossil carbon already present in the atmosphere. It represents the upper limit of a safe-climate zone (300–350 ppm) for human civilization. It is the only target so far proposed that is consistent with the avoidance of runaway global warming. The existential challenge we face is expressed as a simple target figure, first defined by NASA's James Hansen and colleagues in their key 2008 scientific paper, "Target Atmospheric CO2—Where Should Humanity Aim?" which states:

> *If humanity wishes to preserve a planet similar to that on which civilization developed and to which life on Earth is adapted, paleoclimate evidence and ongoing climate change suggest that* CO_2 *will need to be reduced from its current 385 ppm to at most 350*

> *ppm, but likely less than that... An initial 350 ppm CO_2 target may be achievable by phasing out coal use except where CO_2 is captured, and adopting agricultural and forestry practices that sequester carbon. If the present overshoot of this target CO_2 is not brief, there is a possibility of seeding irreversible catastrophic effects.*[1]

We are honored to present here the Dalai Lama's official endorsement of the 350 ppm target. Among the growing list of other international figures supporting it are Nobel Laureate Archbishop Desmond Tutu, Indian environmental leader Dr. Vandana Shiva, Canadian biologist and broadcaster Dr. David Suzuki, Dr. Hermann Scheer, chairman of the World Council for Renewable Energy, and Sheila Watt-Cloutier, chairperson of the Inuit Circumpolar Council.

—Ed.

THE DALAI LAMA

ENDORSEMENT

Right now our greatest responsibility is to undo the damage done by the introduction of fossil carbon dioxide into the atmosphere and climate system during the rise of human civilisation.

We know that we have already exceeded the 350 parts per million that is a safe level of carbon dioxide in the atmosphere. In doing so, we have ushered in a global climate crisis. This is evident from the frequent extreme weather events we witness around us, the unprecedented melting of the Arctic sea-ice and of the great Tibetan glaciers at the Earth's *Third Pole*.

It is now urgent that we take corrective action to ensure a safe climate future for coming generations of human beings and other species. That can be established in perpetuity if we can reduce atmospheric carbon dioxide to 350ppm. Buddhists, concerned people of the world and all people of good heart should be aware of this and act upon it.

December 20, 2008

David R. Loy and John Stanley

What's Next?

One day during meditation, I was contemplating global warming… With some anguish, I asked Nature this question: "Nature, do you think we can rely on you?" I asked the question because I know that Nature is intelligent, she knows how to react, sometimes violently, to re-establish balance. And I heard the answer in the form of another question: "Can I rely on you?" The question was being put back to me: can Nature rely on humans? And after long, deep breathing, I said "Yes, you can mostly rely on me." And then I heard Nature's answer, "Yes, you can also mostly rely on me." That was a very deep conversation I had with Nature.

—*Thich Nhat Hanh*[1]

The human spirit has produced wonderful spiritual, artistic, scientific, and cultural accomplishments, yet the human species as a whole is also "the most dangerous animal" on Earth.[2] We have acquired the capacity to destroy the climate of our own planet. What will we call upon for guidance, as time runs short for corrective action?

In technologically-advanced consumer societies, the rates of anxiety, stress, and mental illness are greater than ever previously recorded.[3] On a physical level, cancer, cardiovascular disease, inflammatory, and auto-immune disease as well as diverse "functional illnesses" have become epidemic. Governments and corporations have decision-making power over the life or death of the biosphere from which our species evolved, but do they even understand the relevant scientific facts? An obsessive focus on the generation of "wealth" has recently provoked a deep crisis in a financial and economic system that must keep growing if it is not to collapse. Always, behind the shop-window of "current events" lies the chaotic decline of the fossil fuel economy.

Looking deeper again, we observe that biological evolution is ecology multiplied by Deep Time. Our species is bringing about the sixth great extinction in Earth's history. General ignorance and indifference about this greatly endangers us too. We used to be one among three human species on Earth—the others, *Homo erectus* and *Homo neanderthalensis*, became extinct. Ninety-nine percent of all species that have ever lived on this planet have become extinct. If we drive through the red-light of climate tipping points, runaway global warming threatens our species with a similar fate. It is what became of most ancient species 55 million years ago, caught in the runaway global hothouse that ended the Palaeocene Epoch.

If we ask why our social evolution has become so maladaptive, we come upon the vivid and hypnotic imagery of television, films, and advertising, a seamless "virtual reality" that now programs our collective nervous system. By manipulating the gnawing sense of lack that haunts our insecure sense of self, the *attention economy* insinuates its basic message deep into our awareness: the solution to any discomfort we might have is consumption.[4] Needless to say, this all-pervasive conditioning is incompatible with the liberative path of Buddhism. What does Buddhism have to offer when our collective situation seems to have become so intractable?

Buddhism has powerful cultural assets. It has long-established ethical teachings and contemplative practices, the moral authority of traditional religious communities, and the potential political power of at least 376 million adherents. Above all, Buddhism is based on the recognition of *interdependence*, the spiritual truth that biologists have also elucidated through the scientific disciplines of ecology, evolution, and molecular genetics. And whether we like it or not, we have entered the century of the environment. In this century, Buddhism has a special destiny. There has never been a more important time in history to bring all the resources of Buddhism together, on behalf of all sentient beings. There has never been a time when transportation and communication systems make this as possible as they do now. Buddhist spiritual power could create examples of change that influence the whole world.

Many Buddhist public events, rituals, and projects are traditionally dedicated to world peace. However, environmental catastrophes and struggles over fossil fuels are already making world peace impossible. According to the United Nations, sixty nations, most of them economically undeveloped and many politically unstable, will experience grave difficulties amplified by ever-scarcer resources and climate chaos. Even countries not directly affected are likely to be flooded by millions of refugees. These are very practical matters for the survival of Buddhism in Asia, as well as for world peace generally.

We have a brief window of opportunity to take action, to ensure the continuity of the many varied and beautiful forms of life on Earth, including our own. Future generations, and the other species that constitute the living world, have no voice to ask for our compassion, wisdom, and leadership. We must listen to their silence. We must be their voice, too, and act on their behalf.

Notes on Translations

a The dynasty of the kings of Ayodhya, descended from Iksvaku, a son of the primordial divine offspring, Manu, into whose lineage "universal monarchs" (*cakravartin*), including Buddha Shakyamuni, are said to have been born.

b One of the five wish-granting trees of Indra's paradise, and a metaphor for any productive and bountiful source.

c Jambudvipa ("rose apple" continent), the southern continent of traditional Buddhist cosmology, where human beings are said to reside in an environment that is more conducive to spiritual practice.

d The Buddha.

e Guru Nangsi Zilnon is a resplendent form of Guru Rinpoche.

References

Introduction

The Buddhadharma and the Planetary Crisis

1. Thich Nhat Hanh, *The Art of Power* (New York: HarperOne, 2008).
2. David Spratt and Philip Sutton, *Climate Code Red* (Melbourne, Australia: Scribe Publications, 2008).
3. Dalai Lama XIV, *His Holiness the XIV Dalai Lama on Environment—Collected Statements* (Dharamsala: DIIR Publications, 2007).
4. Robert Jensen, "The Delusion Revolution: We're on the Road to Extinction and in Denial," August 2008, AlterNet.org
5. Translated by David Karma Choephel, 2007.

Global Warming Science—A Buddhist Approach

Climate, Science, and Buddhism

1. Dalai Lama XIV, *The Universe in a Single Atom* (New York: Broadway, 2006).

Our Own Geological Epoch: The Anthropocene

1. Sakya Trizin Rinpoche, "The Global Ecological Crisis: An Aspirational Prayer." In this volume.
2. Mark Lynas, *Six Degrees—Our Future on a Hotter Planet* (London: Harper Perennial, 2008).

3. United Nations Environment Programme, *Global Environment Outlook 4 (GEO-4): Summary for Decision Makers*, 2007, www.unep.org
4. Intergovernmental Panel on Climate Change, *IPCC Fourth Assessment Report*, 2007, www.ipcc.ch/ipccreports/assessments-reports.htm
5. Paul Crutzen and Eugene Stoermer, *The "Anthropocene"* (Global Change Newsletter 41, 2000).
6. Tim Flannery, *The Weather Makers* (New York: Atlantic Monthly Press, 2006).
7. David Spratt and Philip Sutton, *Climate Code Red.*

"The celestial order disrupted loosens plague, famine, and war..."

1. Dalai Lama XIV, "Universal Responsibility and the Climate Emergency." In this volume.
2. Dalai Lama XIV, *His Holiness the XIV Dalai Lama on Environment—Collected Statements.*
3. Intergovernmental Panel on Climate Change, *IPCC Fourth Assessment Report.*
4. Padmasambhava, trans. Keith Dowman, *Legend of the Great Stupa* (Cazadero, CA: Dharma Publishing, 1973).
5. David Spratt and Philip Sutton, *Climate Code Red.*
6. Robert Nicholls, et al., *Climate Change Could Triple Population at Risk from Coastal Flooding by 2070, Finds OECD*, April 2007, www.oecd.org

Climate Breakdown at the Third Pole: Tibet

1. Sakya Trizin Rinpoche, "The Global Ecological Crisis: An Aspirational Prayer." In this volume.
2. Jane Qiu, "The Third Pole," *Nature* 454 (2008).

3. David Spratt and Philip Sutton, *Climate Code Red.*
4. Lester Brown, *Plan B 3.0* (New York: W.W. Norton, 2008).

The Road from Denial to Agricultural Collapse

1. Dzongsar Khyentse Rinpoche, "A Prayer to Protect the World's Environment." In this volume.
2. Intergovernmental Panel on Climate Change, *IPCC Fourth Assessment Report.*

The Peak Oil Factor

1. Dilip Hiro, *Blood of the Earth* (New York: Nation Books, 2006).
2. David Strahan, *The Last Oil Shock* (London: John Murray, 2007).
3. Commission on Oil Independence, *Making Sweden an Oil-Free Society*, Prime Minister's Office, Sweden, June 2006, www.sweden.gov.se
4. Richard Heinberg, *The Oil Depletion Protocol* (Gabriola Island, BC: New Society Publishers, 2006).
5. Dzigar Kongtrul, "Minimum Needs and Maximum Contentment." In this volume.

Scientific Predictions of Ecological Karma

1. Dalai Lama XIV, "Universal Responsibility and the Climate Emergency." In this volume.
2. Ed Ayres, *God's Last Offer* (New York: Four Walls Eight Windows, 1999).
3. The Global Footprint Network: Advancing the Science of Sustainability, www.footprintnetwork.org
4. Mark Lynas, *Six Degrees.*
5. www.rainforestconcern.org
6. Timothy Lenton, "Tipping Points in the Earth's Climate System" (research paper, University of East Anglia, 2007), www.researchpages.net

7. James Hansen, "Global Warming: Is There Still Time to Avoid Disastrous Human-Made Climate Change?" (presentation, National Academy of Sciences, Washington, DC, April 2006), www.columbia.edu/~jeh1
8. James Hansen, "Global Warming Twenty Years Later: Tipping Points Near" (briefing, Select Committee on Energy Independence and Global Warming, U.S. House of Representatives, June 2008), www.columbia.edu/~jeh1

The Sixth Great Extinction

1. Edward Wilson, *The Future of Life* (New York: Vintage, 2003).
2. Tim Flannery, *The Weather Makers.*
3. Dalai Lama XIV, *His Holiness the XIV Dalai Lama on Environment—Collected Statements.*
4. Robert Jensen, "The Delusion Revolution: We're on the Road to Extinction and in Denial."
5. John Daido Loori, *Teachings of the Earth* (Boston: Shambhala Publications 2007).

What Makes Us Do It?

1. William Blake, "The Human Abstract," *Songs of Innocence and of Experience* (Oxford: Oxford University Press, 1970).
2. Jared Diamond, *The Rise and Fall of the Third Chimpanzee* (New York: Vintage, 1992).
3. Robin Dunbar, *The Human Story* (London: Faber and Faber, 2005).
4. Matt Ridley, *The Origins of Virtue* (New York: Penguin, 1998).
5. Thich Nhat Hanh, *The World We Have* (Berkeley: Parallax Press, 2008).
6. James Hansen, "Global Warming: Is There Still Time to Avoid Disastrous Human-Made Climate Change?"
7. Dalai Lama XIV, *His Holiness the XIV Dalai Lama on Environment—Collected Statements.*

A Safe-Climate Future

1. David Spratt and Philip Sutton, *Climate Code Red.*
2. James Hansen, "Dear Michelle and Barack" (open letter, December 2008), www.columbia.edu/~jeh1

Asian Buddhist Perspectives

Preface—The Meaning of Aspirational Prayer

1. Franklin Edgerton, *Buddhist Hybrid Sanskrit Dictionary* (Delhi: Motilal Banarsidass, 1998).
2. Bhikkhu Bodhi, trans., *The Connected Discourses of the Buddha* (Boston: Wisdom Publications, 2002).
3. Har Dayal, *The Bodhisattva Doctrine in Sanskrit Buddhist Literature* (Delhi: Motilal Banarsidass, 1999).
4. Thomas Cleary, trans., *The Flower Ornament Scripture* (Boston: Shambhala Publications, 1993).
5. Dudjom Rinpoche, *The Nyingma School of Tibetan Buddhism* (Boston: Wisdom Publications, 2002).

Western Buddhist Perspectives

The Voice of the Golden Goose

1. Spencer Weart, *The Discovery of Global Warming* (Cambridge, MA and London: Harvard University Press, 2008).
2. Joseph Romm, *Hell and High Water: Global Warming—the Solution and the Politics—and What We Should Do About It* (New York: William Morrow, 2007).
3. James Hansen, "Global Warming Twenty Years Later: Tipping Points Near."
4. Emily Robinson, "Exxon Exposed," *Catalyst: The Magazine of the Union of Concerned Scientists* (Spring 2007).
5. Bernie Sanders, "Global Warming Is Reversible," *The Nation*, November 2007, www.thenation.com

6. Jake Schmidt, "Detailed Actions to Restore America's Global Leadership on Global Warming," *The Environmentalist Magazine*, November 2008, www.the-environmentalist.org

Now the Whole Planet Has Its Head on Fire

1. Garma C.C. Chang, *The Buddhist Teaching of Totality: The Philosophy of Hwa Yen Buddhism* (State College, PA: Pennsylvania State University Press, 1971).
2. Thomas Cleary, *Entry Into the Inconceivable: An Introduction to Hua-yen Buddhism* (Honolulu: University of Hawaii Press, 1995).
3. Stephanie Kaza, "To Save All Beings," in *Engaged Buddhism in the West*, Christopher Queen, ed. (Boston: Wisdom Publications, 2000).
4. Shohaku Okumura, Taigen Dan Leighton, *The Wholehearted Way: A Translation of Eihei Dogen's* Bendowa (Boston: Tuttle Publishing, 1997).
5. Donald Mitchell, *Buddhism: Introducing the Buddhist Experience* (Oxford: Oxford University Press, 2007).
6. Joanna Macy, *Dharma and Development* (Sterling, VA: Kumarian Press, 1991).
7. Joanna Macy, *World as Lover, World as Self* (Berkeley, CA: Parallax Press, 1991).
8. Naomi Klein, *The Shock Doctrine: The Rise of Disaster Capitalism* (New York: Picador, 2008).
9. Gary Snyder, *A Place in Space: Ethics, Aesthetics, and Watersheds* (Berkeley, CA: Counterpoint, 2008).
10. Donald Rothberg, *The Engaged Spiritual Life* (Boston: Beacon Press, 2006).

The Future Doesn't Hurt... Yet

1. Independent review, "Stern Review on the Economics of Climate Change," HM Treasury, Great Britain, 2006, www.hm-treasury.gov.uk/stern_review_report.htm

Solutions

Clarity, Acceptance, Altruism—Beyond the Climate of Denial

1. David Livingstone Smith, *Why We Lie* (New York: St. Martin's Griffin, 2007).
2. Tim Flannery, *The Weather Makers*.
3. Kevin Philips, *American Theocracy* (New York: Viking Adult, 2006).
4. Al Gore, *An Inconvenient Truth* (New York: Rodale Books, 2006).
5. Gabrielle Walker and Sir David King, *The Hot Topic* (New York: Mariner Books, 2008).
6. Energy Watch Group, "Wind Power in Context—A Clean Revolution in the Energy Sector," 2008, www.energygroup.org
7. Thich Nhat Hanh, *The World We Have*.
8. Bruce Levine, *Surviving America's Depression Epidemic* (White River Junction, VT: Chelsea Green Publishing, 2007).
9. D.T. Suzuki, *Zen and Japanese Culture* (Princeton, NJ: Princeton University Press, 1970).

A Renewable Future

1. Robert Aitken, *The Morning Star* (Berkeley, CA: Shoemaker and Hoard, 2003).
2. Richard Heinberg, *The Party's Over* (Gabriola Island, BC: New Society Publishers, 2005).
3. Dilip Hiro, *Blood of the Earth*.
4. Herman Scheer, *Energy Autonomy* (London: Earthscan Publications, 2007).

5. Ken Zweibel, James Mason, and Vasilis Fthenakis, "A Solar Grand Plan," *Scientific American*, December 2007, www.sciam.com
6. Lester Brown, *Plan B 3.0.*
7. Edward Wilson, *The Future of Life.*

Five Transformative Powers

1. Lester Brown, *Plan B 3.0.*
2. General Compression: Wind Energy on Demand, www.generalcompression.com
3. Massachusetts Institute of Technology, "The Future of Geothermal Energy," U.S. Department of Energy, 2006, www.eere.energy.gov

The End of Energy Waste

1. Charles F. Kutscher, ed., "Tackling Climate Change in the U.S.," American Solar Energy Society, January 2007, www.ases.org
2. John Rennie, ed. "Energy's Future: Beyond Carbon," special issue, *Scientific American*, September 2006.

Goodbye to the Internal Combustion Engine

1. Lester Brown, *Plan B 3.0.*
2. Motor Development International, www.mdi.lu

Tradable Energy Quotas

1. Tradable Energy Quotas, www.teqs.net
2. David Strahan, *The Last Oil Shock.*

Drawing Down Carbon with Biochar

1. Anne Casselman, "Inspired by Ancient Amazonians, a Plan to Convert Trash into Environmental Treasure," *Scientific American*, May 2007, www.sciam.com

Reducing the Carbon Footprint of the Meat Industry

1. Food and Agriculture Organization of the United Nations, "Livestock's Long Shadow: Environmental Issues and Options," Rome, 2007, www.fao.org
2. World Cancer Research Fund / American Institute for Cancer Research, *Food, Nutrition, Physical Activity, and the Prevention of Cancer: A Global Perspective* (Washington, DC: AICR, 2007).

Ending Deforestation

1. Raymond Gullison et al, "Tropical Forests and Climate Policy," *Science*, vol. 316, no. 5827 (2007).

Reforesting the Earth

1. Dalai Lama XIV, "The Sheltering Tree of Interdependence." In this volume.
2. Lester Brown, *Plan B 3.0.*
3. David Spratt and Philip Sutton, *Climate Code Red.*
4. The United Nations Environment Programme's Billion Tree Campaign, www.unep.org/billiontreecampaign
5. Trees for Life: Restoring the Caledonian Forest, www.treesforlife.org.uk
6. Al Gore, *Earth in the Balance* (New York: Rodale Books, 2006).
7. Gyalwang Karmapa XVII (this volume).

Afterword

Endorsement of a Safe Level for Atmospheric Carbon Dioxide

1. James Hansen et al, "Target Atmospheric CO_2—Where Should Humanity Aim?" *Open Atmospheric Science Journal*, vol. 2 (2008), pp. 217–231.

What's Next?

1. Thich Nhat Hanh, *The Art of Power.*
2. David Livingstone Smith, *The Most Dangerous Animal* (New York: St. Martin's Griffin, 2007).
3. Bruce Levine, *Surviving America's Depression Epidemic.*
4. David R. Loy, *Money, Sex, War, Karma* (Boston: Wisdom Publications, 2008).

About the Editors

John Stanley, Ph.D., is a biologist who has led university and governmental research groups in university and government in Canada, Switzerland, and the United Kingdom. His doctoral and postdoctoral research was in the genetics of symbiotic nitrogen fixation. He has published many scientific papers in the fields of molecular genetics, microbiology, and public health, and is a member of the New York Academy of Sciences. He is a practicing member of the International College of Applied Kinesiology, with a longstanding personal interest in functional medicine. He and his wife Diane have been practitioners of Tibetan Buddhism in the Nyingma tradition of Dudjom Rinpoche for three decades, and are based in western Ireland.

David R. Loy, Ph.D., is Besl Professor of Ethics/Religion and Society at Xavier University in Cincinnati, Ohio. His work is primarily in comparative philosophy and religion, particularly addressing Buddhism in the modern world. His books include *Nonduality: A Study in Comparative Philosophy*; *Lack and Transcendence: The Problem of Death and Life in Psychotherapy*; *Existentialism and Buddhism*; *A Buddhist History of the West: Studies in Lack*; *The Great Awakening: A Buddhist Social Theory*; *Money, Sex, War, Karma: Notes for a Buddhist Revolution*; and *Awareness Bound and Unbound: Buddhist Essays*. A Zen practitioner for many years, he is qualified as a teacher in the Sanbo Kyodan tradition of Japanese Buddhism.

Gyurme Dorje, Ph.D., holds higher degrees in Tibetan literature, Sanskrit, and Oriental Studies. He is a leading scholar of the Nyingma tradition of Tibetan Buddhism and has established a small research library for Tibetan studies in Perthshire, Scotland. He is the director of Trans Himalaya expeditions and cultural tours (www.trans-himalaya.com). His nine major publications include works on Tibetan lexicography, medicine, and divination, pilgrimage guides to Tibet and Bhutan, translations of the late Kyabje Dudjom Rinpoche's great works contained in *The Nyingma School of Tibetan Buddhism: Its Fundamentals and History*, and the first complete translation of *The Tibetan Book of the Dead*.

About Wisdom Publications

Wisdom Publications, a nonprofit publisher, is dedicated to making available authentic works relating to Buddhism for the benefit of all. We publish books by ancient and modern masters in all traditions of Buddhism, translations of important texts, and original scholarship. Additionally, we offer books that explore East-West themes unfolding as traditional Buddhism encounters our modern culture in all its aspects. Our titles are published with the appreciation of Buddhism as a living philosophy, and with the special commitment to preserve and transmit important works from Buddhism's many traditions.

To learn more about Wisdom, or to browse books online, visit our website at www.wisdompubs.org.

You may request a copy of our catalog online or by writing to this address:

Wisdom Publications
199 Elm Street
Somerville, Massachusetts 02144 USA
Telephone: 617-776-7416
Fax: 617-776-7841
Email: info@wisdompubs.org
www.wisdompubs.org

The Wisdom Trust

As a nonprofit publisher, Wisdom is dedicated to the publication of Dharma books for the benefit of all sentient beings and dependent upon the kindness and generosity of sponsors in order to do so. If you would like to make a donation to Wisdom, you may do so through our website or our Somerville office. If you would like to help sponsor the publication of a book, please write or email us at the address above.

Thank you.

Wisdom is a nonprofit, charitable 501(c)(3) organization affiliated with the Foundation for the Preservation of the Mahayana Tradition (FPMT).